THE SAH

Population

Integrated Rural Development Projects

Research Components in Development Projects

Proceedings of the 6th Danish Sahel Workshop, 6—8 January 1994

Edited by Annette Reenberg & Birgitte Markussen

AAU REPORTS 32

Department of Systematic Botany, Aarhus University

This issue was sponsored by
Danida

Editors

Anette Reenberg. *Born 1948. Since 1974 associate professor at Institute of Geography, University of Copenhagen. Research areas: Ecological geography with emphasis on landscape ecology and agricultural systems (agricultural landscapes in Denmark; agricultural systems, land use and environment in the Sahel). Address: Institute of Geography, Øster Voldgade 10, DK-1350 Copenhagen K, Denmark.*

Birgitte Markussen. *Born 1963. M. A. student at Department of Ethnography and Social Anthropology, Aarhus University. Specializing in development support communication and educational video production. Diploma in video production from VARAN - Ateliers de Réalisation Cinémagraphique financed by the French Ministry of Foreign Affairs. Present address: Carl Bernhards Vej 3, DK-1817 Frederiksberg C, Denmark.*

Contents

Part III: Research Components in Development Projects

Introduction

The present volume contains the contributions to the Danish Sahel Workshop held in Sønderborg 6-8 January 1994. This workshop is number 6, and the arrangement has thereby developed to what you might call a well established yearly tradition.

During these six years, the workshop has provided an opportunity to meet informally other members of the "Danish Sahelian Resource Base", e.g., people from the research-community, experts working on development projects and representatives from the administration of developmental aid.

It is generally considered that the workshop has been a fruitful forum for discussions and exchange of views. We therefore thank Danida and the people supporting the idea of the workshop in Danida for the funding which makes this arrangement possible. The generous support for the last couple of years enabled us to invite colleagues from the Sahelian region and from strong European research groups to contribute to our workshop with their ideas and insight in the Sahelian problems.

The main themes for this year workshop largely followed the recommendations given last year, and the focus has been on three relatively independent issues:

* In PART I, the question will be addressed: Is the Sahel overpopulated? Speakers from different disciplines ranging from natural science to social and economic sciences were asked to give their views and evaluate the population pressure in the Sahel.

* In PART II, integrated rural development projects were brought into focus. The aim of the presentations is to discuss what type of empirical information is needed as a basis for this type of project.

*   Finally, in PART III, research components in development projects are discussed and evaluated in the light of Scandinavian experiences.

The workshop has been arranged by an organizing committee consisting of Tove Degnbol (International Development Studies, RUC), Kristine Juul (International Development Studies, RUC), Ivan Nielsen (Institute of Biological Sciences, AU), Christa Nedergaard Rasmussen (Danish Red Cross), and the editors of this volume.

We would like to thank the workshop speakers for contributing the papers. We gratefully acknowledge the very careful and comprehensive linguistic correction of the English contributions made by Dr. Robbin Moran (Institute of Biological Sciences, AU). We would also like to express our gratitude to Patricia Bussone for her correction of the French papers. The papers have not been scientifically reviewed and are in principle printed as presented by the authors with editorial corrections only.

Copenhagen, March 1994.

Anette Reenberg
Birgitte Markussen

Population Density, Carrying Capacity and Agricultural Production Technology in the Sahel

Leo Stroosnijder
Department of Irrigation and Soil & Water Conservation
Wageningen University, The Netherlands

Introduction

The Sahel is among the most sparsely populated areas of the world. But it is also believed that this area is not being used in a sustainable manner. It faces loss of biodiversity and is subject to degradation, desertification and thus heading towards environmental disaster.

The prospect of a doubling of the population over the next twenty years has raised a number of questions, such as: Is the increase in population density to blame for the present-day environmental degradation? And: If the answer to the first question is yes, then will this positive correlation between population growth and environmental degradation continue? In search of an answer to the first question, this one is often asked differently in the form of: Is the present population density in the Sahel beyond the carrying capacity of the natural resources? With carrying capacity defined in the traditional sense, only considering renewable resources, the answer to this latter question is certainly yes.

This author, questions the relevance of the carrying capacity concept on ecological, agricultural, and ethical grounds. This paper says farewell to the carrying capacity concept.

Instead we raise the following question: Does the present population apply a production technology that uses natural resources in a sustainable way? We present a broad overview of the technology that is used at present and of the signs of degradation. This leads to the conclusion that technology has not changed fast enough to match population growth. One must develop new land use systems by matching improved agricultural technology with the natural resources while simultaneously solving a number of economic, social, and political issues. But what are the potential of these natural resources? We present a brief overview of the natural resources, not as usual from a renewable resources perspective but from a technology perspective, and we introduce five developmental steps that may trigger the technological change that is needed to develop sustainable farming systems in the Sahel. New tools like Quantified Land Use Planning (QLUP) and Quantified Farming Systems Analysis (QFSA) can be of help. But another question comes up: Might the present, low population density trigger development that is a prerequisite to be able to change agricultural production practices by applying new technology? In the conclusion, we answer the original two questions.

Farewell to the Carrying Capacity Concept

Carrying capacity (CC) can be defined as: the maximum level of sustainable exploitation of the renewable resources (after Kessler 1994). In the largest part of the Sahel the most limiting renewable resource is the natural fertility (Nitrogen and Phosphorus) of the soil. Hence, 'sustainable' in the above definition means that an agricultural practice may not deplete the stock of plant nutrients in the soil. The word 'maximum', in the above definition, refers to production situations without losses of that part of the renewable resources that becomes available each growing season for agricultural production. Following this definition, the CC can be considered as a potential property of the ecosystem only, regardless of the agricultural technology that is used.

From the signs of over-exploitation, one may not conclude that the present population density is above the CC. After all, the CC is a potential value not reflecting losses of water and nutrients and especially in the Sahel, due to the harsh climatic conditions, such losses are very common. The actual-CC is much lower than the potential-CC, and population density can be above this actual-CC but be still below the potential-CC.

Before assessing the relevance of the Carrying Capacity concept, we define two major types of land use in the Sahel: (1) animal husbandry through pastoralism and nomadism and (2) arable and mixed farming. The former land use can be considered as the use of natural ecosystems with little human interference and hardly any external inputs used. Arable and mixed farming range from the use of zero inputs to semi-intensive farming using medium levels of inputs.

First, the usefulness of the CC-concepts for ecosystems with almost no human interference will be discussed. The CC-concept was first used by wild-life ecologists to estimate the sustainable density wild herbivores. The CC-concept was an extension of the continuity principle of older theories like gradual retrogression and secondary succession. For extensive grazing in the Sahel, the CC-concept is considered to be too simple by Leeuw and Tothill (1990) and later also by Behnke and Scoones (1992). For semi-arid regions with large spatio-temporal variations, the so-called state-and-transition models and the concept of thresholds seem more appropriate (Friedel 1991). Lately, properties of semi-arid rangelands like modality, discontinuity, divergence and hysteresis have lead to an interest in the catastrophic theory of Thom (1978), referring to the stability and resilience of ecosystems. So, the CC-concept is hardly used any more for natural ecosystems (Rietkerk 1994).

Neither for agricultural land use, the CC-concept will not be successful in terms of development prospects. First, the CC deals only with renewable resources. These are poor in the Sahel, so that only low population densities can be maintained. Breman (1992) calculated for arable farming in dry years in the southern Sahel a CC of 10 persons/km^2 and for the Sudanian zone a CC of 36. If compared with present average population densities of 13 and 33 respectively, this implies overexploitation of the renewable resources. This is not surprising because it occurred in previous times in many parts of the world when agriculture was based on renewable resources only.

The second reason is that the CC-concept gives a hypothetical potential value. The gap between actual-CC and potential-CC can only be narrowed by the application of agricultural technology. For instance, with simple local soil and water conservation techniques (without inputs from outside) losses of water and nutrients can be reduced considerably so that the actual CC will increase. Therefore, the potential-CC will be of little practical value. Instead, it is better to link the sustainable output of an agrosystem with the type of technology that is used in that agrosystem.

The third reason is that the use of the CC-concept stresses that the population over-exploits their natural resources and has grown too fast. These conclusions imply blame to the local people which is not justified. No where in the world are people able to sustain their lifestyle based on the sustainable use of only the renewable resources.

There is a certain preference to apply the carrying capacity concept to agriculture in the Sahel. But where in the world are we still taking only the renewable resources into account? In addition to the renewable resources, any agrosystem makes use of production means like labour, capital, and knowledge. Therefore, the production level that can be reached, and hence the number of people that can be feed without destroying the natural resource base, is everywhere more a function of the production technology used than of the renewable resources!

Hence, there are several reasons to say farewell to the CC-concept both for the assessment of ecosystems as well as for the productivity of agrosystems. The production technology used, even in so-called closed systems, has a large influence on the actual-CC. A potential-CC value therefore loses its practical use because it is not linked to a certain production technology. Technology transfer and development is the key to reduce losses of renewable resources. And similar technology can also be used to add inputs to the agrosystem. There is no benefit in creating a strict boundary between renewable resources and inputs. The potential-CC has too long and too much been associated with no or low input agriculture and has stressed the difference between renewable and non-renewable resources. The latter have been stigmatized due to the supposed reliance of fossil energy for their production.

Newer concepts such as sustainability, agricultural intensification (Keulen and Breman 1990), Quantified Land Use Planning (QLUP: Fresco et al. 1994) and Quantified Farming Systems Analysis (QFSA: Stroosnijder et al. 1994) are more suitable to the discussion of how land-use systems can be developed which are in equilibrium with the population density. Provided that the complexity of the sustainability concept is recognized (Jones 1993) and that time and spatial scales are properly distinguished (Fresco and Kroonenberg 1992). Sustainable land use (in semi-arid zones) is defined as: The maintenance or improvement (over several years of fluctuating rainfall) of soil chemical, physical, and biological properties with respect to a certain baseline. Agricultural intensification is defined as: an increase in inputs of labour, capital or knowledge, for the purpose of increasing the value of output per hectare (after Tiffen et al. 1993). In applying these concepts a clear distinction should be made between agrosystems (where agricultural intensification is the development path) and ecosystems (where nature conservation is the prime development goal). Careful quantitative land-use planning should designate areas for both types of land use. Since the aims for such different land-use types vary considerably, no attempt should be made to compromise in aims or land use (Prins 1992).

Failing Agricultural Technology

The rapid increase in the Sahelian population influences the agricultural production systems. An example is how grain was produced in the past versus the present. Not so long ago, there were distinct ethnic groups of farmers and herders. Mixed farming, where one household is occupied with arable farming as well as with animal husbandry, was rare. Farmers grew grain in a shifting cultivation land use system. Fields were not abandoned due to weed infestation (as in the humid tropics) but due to a decrease in fertility. In normal rainfall years, yields were abundant with respect to the labour investment. So, it was relatively easy to keep grain stocked for 2-3 years which is needed in areas with periodic sequences of drought years.

When population density increased, fallow periods in the shifting cultivation system decreased and production in kg/ha decreased. As a result, the number of hectares needed for subsistence increased and labour productivity (hours per kg produced) decreased to a level where food production is limited by labour shortage during soil preparation and sowing which are the peaks in labour demand. It was no longer easy to produce excess food in normal rainfall years so that food security deteriorated due to smaller stocks kept for dry years.

Farmers were well aware of the cause of the deterioration, i.e. nutrient depletion of the soil. They responded two ways. They became interested in animal husbandry for the prime reason to move nutrients (in the form of manure) from the common grounds (brousse) to their fields. Fertilization of the fields in the direct neighbourhood of their dwellings started and so-called 'champs de cases' were born where the drop in yields could be halted at least. On the other hand, they expanded (with animal traction if possible) their area under shifting cultivation in the common 'brousse' further away from the village.

So, total agricultural production tried to follow population growth mainly by an expansion of the area under cultivation. An expansion which was made possible by the sparseness of the absolute population density. Under these conditions the agricultural production technology (use of inputs, cultural practices, etc.) hardly changed in recent decades. This is partly due to a failing extension system. The proposed technological packages in rural extension have virtually not changed, whereas the farmer's social and material conditions have worsened. It is also true that in most cases, extension workers bring along a standard package that disregards different farming systems. Only recently it was discovered by the extension services that the farming community is not a homogeneous entity, but consists of a wide variation of farm types, ecological circumstances, social and cultural groups, and that farmers can be females as well.

A direct agricultural effect of the unadapted technology and the decrease of yields per hectare is the cultivation of much land marginal. In fact, an extensification of arable farming has taken place. Less and less time is available to take care of the land. Under such conditions, erosion increases and causes environmental degradation. In the Sahel, there are obvious signs of degradation such as the loss of productivity by failure to replace nutrients or the modification of the soil's physical properties. The apparent over-exploitation leads to soil mining (Pol 1992), chemical and physical degradation due to a depletion of the organic matter content of the soil (Stroosnijder 1992) and increased runoff and erosion.

An ecological effect is the decrease of vegetation cover, standing biomass, and soil organic matter. The majority of plant nutrients are stored in this organic matter. Each year about 2% of the available organic matter mineralizes so that Nitrogen (N) and Phosphorus (P) become available (Pieri 1990). In dynamic equilibrium, the formation of soil organic matter is equal to the losses leading to a constant level of soil organic matter. Fresh organic material must be added to the soil each year and converted into organic matter by microbes in the soil. Over generations, farmers were able to maintain this renewable resource base by utilizing pastoral animal husbandry systems and shifting agriculture for crop production. In recent decades, biomass losses have increased rapidly. The losses are due to the breakdown of dried or dead biomass, occasional burning and the use of biomass by man and animals. In the opening-up of new land for arable farming, huge amounts of biomass were burned and the organic matter in the top-soil has been carried away by water and wind. Man started burning the savannah as a management practice for animal husbandry so that much Carbon (C) and Nitrogen (N) were lost.

The resulting decrease in organic matter content of the soil means that less plant nutrients become available each year, triggering a downward spiral in biomass production. In addition, lower soil organic matter influences the physical properties of the soil. Soil structure decreases and surface crusting increases. The fraction of the annual rainfall that does not infiltrate into the soil but runs off into depressions or rivers can increase to 60%. This further aggravates the downward spiral in biomass production (Stroosnijder 1994).

The decreasing quality of the living conditions also resulted in social change. Only recently research is started looking into stress and acculturation due to environmental degradation (van Haaften, pers. comm.). Studies of the local social structures are hampered by the already long period of interference by many different help organisations (Lekanne, pers. comm). At present, one may wonder what has damaged the local social institutions more, drought and degradation or help organisations. For instance, the more than 150 institutions operating at the central Mossi Plateau in Burkina Faso have resulted in opportunism towards the unpredictable policies of numerous competing help organisations as the dominant social mechanism.

Natural Resources Perspectives

The Sahel zone in West-Africa covers about 5,400,000 sq. kms. and harboured about 42 millions inhabitants in 1990. Agriculture is the main source of income in the Sahel. Over 80% of the population is working in this sector; on the average it provides for 40% of the gross domestic product, and in some countries up to 60%. Only 12% of the total surface are rainfed agricultural lands with soils of a low fertility. Agriculture is heavily based on food production, and food security is the main preoccupation of most Sahelian countries. Agricultural production during last decades did not match the population growth, and most countries are not self-sufficient in food production (World Bank 1989).

Soils
In the Sahel (100-600 mm) and Sudan zone (600-1200 mm) three major landscape units are distinguished: Sandy Complex, Detritic Complex on Sandstone or Laterite and Fluviatile or lacustrine Complex (Stroosnijder 1982).

Sandy Complex (50% in Sahel, 10% in Sudan zone): e.g. eolian deposits, deep and often uniform over vast regions dominating the north Sahel. Soils are sandy or loamy sand, and rarely, sandy loam. The soil surface is often sandy enough to maintain a high infiltration capacity so that runoff is almost absent. Erosion by water and wind is limited except in the northern Sahel. The vegetation on this complex is the most homogeneous of the three. Soil depends on the age of the deposits. Yellow sand is older than red sand and hence less fertile.

Detritic Complex on Sandstone or Laterite (30% in Sahel and 70% in Sudan zone): e.g. soils developed on sandstone or laterite. They may be deep, but rarely are. There are often bare stretches with sandstone or hard laterite on the surface. Most of the soils are loamy, but very heterogeneous, sometimes showing heavy runoff. This is reflected in the vegetation which is often irregular composed of dense bushes and bare spots. There is widespread runoff due to poor infiltration capacity of the loamy soils susceptible to crusting. There is adequate natural drainage and runoff forms temporary pools, flood-plains or flows into wadis. There is much erosion by water. Fertility usually is extremely low. With the runoff there is a permanent export of fertility.

Fluviatile or lacustrine Complex (20% in Sahel and Sudan zone): e.g. soils formed on recently-formed or fossil fluviatile or lacustrine sediments. The deeper layers are clayloam often covered by a layer of eolian sandy loam. Soils may be heterogeneous in a landscape that is generally flat. Natural drainage is hampered so that water stagnates. This complex often receives run-on water from the adjacent ditritic complex. Fertility varies over short distances. Recent soils may be rich, while fossil soils are impoverished. Soils receiving runon with sediment may be regularly enriched.

From the renewable resource viewpoint the soil resource base is old, well weathered with a low cation exchange capacity, hence with low natural fertility. From a technology perspective the situation is much better. There are huge reserves of physically good soils, with adequate topography and internal drainage that can be exploited under agriculture with adapted technology. Little management problems occur, salinity for instance is a minor problem.

Water

There is still no consensus about whether the Sahel is subject to climatic change. Recently, Tiffen et al. (1993) analyzing 100 years (1890-1990) of rainfall in Kenya did not find evidence for climatic change in East Africa. However, clear longterm cycles of dryer and wetter years can be distinguished. Sivakumar (1992) proved climatic change using 30 years (1960-1990) of data for Niger in West Africa.

Average total rainfall is abundant during a brief growing season often causing too wet conditions, runoff and erosion. Rainfall is virtually absent during the remainder of the year making agriculture a seasonal affair. Rainfall distribution over the growing season is irregular which makes agriculture risky if the buffer capacity of the soils is too little. The latter is the case in too shallow, too coarse or too heavy soils. Yearly rainfall variations are large as well, leading to large fluctuations in agricultural potential. In addition to this there is tendency for a clustering of dry and wet years aggravating the problem of food shortages. There is a large potential for water conservation and concentration (harvesting), (Stroosnijder & Hoogmoed 1984). However, at present low fertility and inputs levels, it is hardly worth to reduce water losses due to runoff (Stroosnijder et al. 1994).

Climate and Vegetation
The temperature and irradiation are favourable for fast growing, water efficient, C4 plants, while the extreme change of climate from wet to dry prevents the occurrence of persistent pests, diseases or weed infestations.

Due to the semiarid climate, the natural vegetation is highly dynamic in terms of production as well as in species composition (Ridder et al. 1982). Two scales should be distinguished. Sequences of 3-10 years of dryer and wetter (than the average) years occur. This results in north-south shifts or vice versa of the various vegetation zones and certain species of hundreds of kilometres. In addition, inter-annual variation, often at a large spatial scale, makes vegetation look very different from one year to another.

This dynamism has often mislead western scientists visiting the area for a brief period only and have led to many wrong conclusions with respect to the potential (in wetter years) and desertification (in dryer years) of the area. Shifting dunes at the fringe of the Sahara, spectacular as it may look, should certainly not be regarded as desertification.

The natural vegetation of the Sahelian Zone has been remarkable resilient for climatic changes and on a millennium time scale, may be centuries there has been a dynamic equilibrium. However, the present fear is that such a dynamic equilibrium does not exist any more. Non-reversible changes might have occurred, and a threshold might be passed bringing the Sahel in another state.

In summary, in addition to a considerable soil potential, there is ample water for biomass production during the growing season, and temperature and irradiation are favourable for fast growing, water efficient, C4 plant species. In other words, from an agricultural perspective, there is a huge cultivation potential in the Sahel.

How to trigger technological change?

Help organizations can't and should not solve the problems of the Sahel. That should be done by the Sahelian people themselves. However, donors may help the local population both financially and with the transfer of knowledge. In the following, five steps are discussed that are considered crucial for technological change. Each new step follows from the previous one.

Development of local markets for food and technology

At present farmers produce mainly for subsistence. They maintain less reserves for dry years than before because production conditions have worsened. Labour productivity is low and inputs are expensive. Demand at domestic and international markets is low as well as prices are. So, there are no incentives for farmers to produce more than for their subsistence. Low national food security and little income from local trade or export are the results. The only demand market that can and should be developed is the local market. A prerequisite for the development of local markets is a good infrastructure. Therefore, donors should subsidize infrastructural works.

Why is this so important? It is the surplus of their subsistence that farmers should be able to sell at a local market so that the revenues can be used for the application of agricultural production technology. So, development of local markets (demand and supply) are the key for technology change. For the sake of the development of local markets, these should be protected for quite some time (even under GATT) by stopping those imports for which local alternatives become available. Recently, this development policy has been applied with much success by the new Asian tigers.

Neither stimulating production by subsidizing inputs nor price guarantees are viable options as long as the demand for agricultural produce is not increased. In stimulating the execution of infrastructual works one should pay hired labour in cash and stop the institute 'Food for Work'. Terms of trade in the Sahel are at this moment unfavourable for such development because of the local currency policy. This hampers the development of local markets. The currency in the Sahel, the CFA, is much over-valued and the coupling of the CFA with the French FF should be halted.

Diversification

Economic diversification is a prerequisite for the development of local markets for food and technology. Because local demand for agricultural produce can only be generated when other sectors of the local economy can be stimulated to develop. Rural industry, local entrepreneurship and services should be stimulated, and donors can help starting private enterprises with a supply of simple and labour intensive machines.

But also within the agricultural sector one may stimulate a higher degree of job diversification. The dilemma of mechanisation may serve as an example. Mechanisation is only one, out of many, element of the intensification of the agricultural production process. Mechanisation of individual smallholders is seldom feasible due to the smallness of the economic unit and the burden of training too many individuals. And often, especially in the beginning of an intensification process, mechanisation is not desirable. Thus, specialized persons with special equipment should be aimed at. These persons, so-called agricultural contractors, have to make a living (in cash or kind) of their new profession. Stimulating such agricultural private entrepreneurship provides economic diversification of the rural society and adds to a general trend in economic diversification.

Diversification requires a skilled labour force, so schooling becomes important and one should therefore concentrate on vocational training.

Agricultural intensification

With a rising local demand for food, for which a realistic price is paid, increasing agricultural production is a must. After all, with the traditional technology, labour productivity has deteriorated to a level that almost no excess food could be produced. Intensification leads to a higher production per hectare. This implies that enough food can be produced to match the increasing demand, while more areas become available for environmental rehabilitation and protection (Rabbinge 1993). Such intensification implies a higher use of production means like fertilizers and biocides. With higher production a larger part of the yearly rainfall is used for transpiration. This means that gradually it becomes worthwhile to practise water conservation through the use of mechanical measures supported by vegetative measures like lines of perennial grasses and trees.

Intensification of rainfedbased systems increases annual variation in production faster than it increases the average production. This means that annual variations in output will increase. This may lead to large price fluctuations and requires an increase in storage capacity. Where a higher degree of water control is feasible this should be developed as a stabilizing factor on annual agricultural production providing the local population more food security (Huibers and Stroosnijder 1992). There is a large potential for small holders irrigation systems in the inland valleys of the Sahel (Albergel et al. 1993).

Intensification will ultimately lead to a concentration of the agricultural production on a smaller area. This means that the more marginal areas can be used again as silvopastoral areas. These areas serve many functions and are used for extensive grazing, for wood production and the collection of other forest products, for water control and a biological buffer in general (Stroosnijder & Hoogmoed 1993). The present degradation of these areas should be brought under control in order to avoid damage to the adjacent agricultural fields. For certain areas drastic measures such as closing areas for free grazing, as practised in many parts of the world, should be considered. In other areas, biomass production and biomass cover can be stimulated by a stepwise sequential intervention. The latter can often be achieved by simple low costs technology in pace with natural soil recovery. An example of such a package is: first year: stone lines or bunds initiating better establishing conditions for woody species and soil biological activity including termites, second year: stimulation of woody species with local available rock phosphate, fourth year: pruning of woody species for use as cover and stimulation of termites in crusted bare spots between the stone lines or bunds.

Besides the restoration of the silvopastoral areas which will serve only a limited number of pastoralists, there is an urgent need for alternative animal husbandry systems for cattle as well as for shoats. There is a future for mixed farming systems where arable farming and animal husbandry perform synergetic effects (Reijntjes et al. 1992). Therefore, systems with supplementary feeding or with zero grazing should be tested for their viability in mixed farming systems.

In the early stages of intensification, technologies should be introduced that does not or hardly replace labour. Later in the intensification process, when job demand in other sectors of the economy increases, the availability of agricultural labour can become a constraint. Intensification should then be accompanied with the introduction of laboursaving techniques like mechanisation. However, agreement at the private and village level should be obtained about the objectives of the mechanisation because previous introduction of mechanisation did often had an opposite effect. Mechanisation was used to expand the area under extensive cultivation and did not result in intensification.

Land Use Planning

Technology is expensive and financial means are always limited. Inputs are most effectively used in an intensive agricultural production system, in particular when used on the better soils. This stresses the need for quantitative land use planning. Land use planning uses knowledge of natural resources and indigenous knowledge to select appropriate land use scenarios that meet divergent goals of society, local communities and individuals (Fresco et al. 1994). Such land use planning has evolved from a centralized process executed by experts at considerable distance from the field to something quite different: an integrated, interactive process with the attention to its unexpected consequences (Stroosnijder et al. 1994a). Such land use planning must be achieved simultaneously at three different scales: the national level, the regional level and the village and household level. At the national level, the question is where to stimulate what. For instance, whether to spend scarce resources on the development of irrigation or on the development of rainfed agriculture, and which crop to stimulate where. Planning should be specific for each Agro Ecological Zone. Modern technology has become available to apply land use planning at these different levels. Linear Programming and Multi Criteria Analysis allow the aggregation of knowledge at field level at higher levels of integration. Land use planning is a typical interdisciplinary activity (Stroosnijder et al. 1994b).

In order to be able to plan land use, sufficient data must be available. Given the highly dynamic nature of the Sahelian vegetation, monitoring procedures should be improved in order to separate natural short- and long-term changes from irreversible ones. Accessibility of the growing data base should be improved by using a standard Geographical Information System.

Another prerequisite for adequate land use planning is that land rights become assured. The existing repressive legislation in the Sahel often takes away all responsibility for use and misuse of natural resources from the farmer. It is not surprising, therefore, that the farmer, constrained by the socio-economic and ecological situation and having no responsibilities but for his own life, exploits to a maximum the land he is not able to secure on the long run. Securing land rights implies that private smallholdings are stimulated at the expense of communal areas.

Agricultural intensification starts earlier in areas with relative land shortage, i.e. in areas with a high population density (Brouwers 1993 and Tiffen et al. 1993). It might even be a condition, in order to make such intensification working, that the population density has surpassed a certain threshold. Land use planning may help to indicate whether a certain concentration of farming and farmers is needed to surpass a threshold for development and can indicate in which areas such concentrated is most viable.

Research for Development

With donors losing patience with Africa and declining funds for development it is important to stress the need for a long-term political, social, economical and technical commitment. Not only from donors but from the national governments in the first place.

Donors should drastically change their forms of help that created so much local (development) opportunism. Stimulate self-help groups and restore the confidence of people in themselves. Literacy programmes have to be intensified, to assure that at least one member per family can read and write. Experience has shown that crash programmes of six weeks suffice to make an illiterate person literate.

There is a persistent neglect of farmers' knowledge and capability to innovate their farming system. Farmers do have their own experiments without addressing themselves to formal research. Locally developed technologies for breeding of drought resistant sorghum and millet varieties, and animal disease control by medicinal herbs are only a few examples. The dilemma between farmers' and researchers' interests calls for a new approach of agricultural research in the Sahel (Hudson & Cheatle 1993 and Blokland & Stroosnijder 1994). This approach should be based upon mutual respect and open dialogue between farmers, researchers, and development (extension) workers. Also involve private traders and foreign investors of the agricultural sector in the planning and implementation of adaptive research, if possible attract their financial concurrence. Reward researchers for effective field work. A material and moral validation system for practical on-farm research could seduce researchers to stay in the field rather than in the office.

It is important that the local research capacity is enforced and that the research is better linked to development. It is often not a matter of lack of quantity of researchers but of quality. Local research at various levels (including that of the experimenting farmer) should be stimulated in a continuous and coordinated way and only long-term support and commitment should be given (and accepted). This local development should get priority over regional research strengthening. Young researchers should be better trained, develop a research tradition and be more competitive. There certainly is a shortage in national data bases, in documentation centres, in institutional memories. However, it is my own experience that researchers, once provided with literature, find it hard to read and use this literature so that also a change in attitude and research tradition is needed.

Finally, researchers should not be asked to cover the whole information chain from research to extension. Intermediate structures and procedures must be developed that can apply and test research findings in the field before extension takes over. The latter need to be upgraded and enforced in almost all developing countries.

Conclusion

Population growth has indeed caused environmental degradation in the Sahel. Due to an insufficient change in agricultural production technology and the scarceness of the renewable resources, yields per hectare have declined. Counteracting this, land under cultivation has expanded faster than population growth. Similar decreased productivity has occurred on the traditional rangelands. Drought spells did not cause this process but aggravated it and brought it to the forefront.

The observed process is not at all unique for the Sahel. A negative correlation between population growth and environment has been observed in many places in the world where agriculture was at the brink of developing from a low to a medium or high input agribusiness.

By changing the agricultural technology, using more inputs and practising conservation farming, productivity per hectare can be raised quickly and reliably provided that a number of economic, social and political issues can be solved simultaneously. However, inputs must be paid for by production in excess of subsistence that can be sold at a local market. So, in order to stimulate technological change, local markets for food and technology must be stimulated instead of subsidizing inputs or guaranteeing prices. A number of measures are proposed which could trigger technological change. Land use planning can help to determine critical population densities for the intensification of agriculture.

References

Albergel, J., J-M Lamachère, B. Lidon, A.I. Mokadem et W. van Driel (eds.), 1993. Mise en valeur agricole des bas-fonds au Sahel: typologie, fonctionnement hydrologique et potentialités agricoles. Rapport final d'un projet CORAF-R3S, CIEH, Ouagadougou, Burkina Faso, 335 p.

Behnke, R.H. and I. Scoones, 1992. Rethinking Range Ecology: Implications for Rangeland Management in Africa. IIED Paper no. 33. London, UK.

Blokland, A. and L. Stroosnijder, 1994. The Sahel region: Sustainable Agriculture and Food Security: the common challenge of Farmers, Research and Development. In: Development orientated research; a second look at the role of The Netherlands. Groningen Conference, 1992.

Breman, H., 1992. Agro-ecological zones in the Sahel: Potentials and Constraints. P. 19-37 in: Poverty and Development: Analysis & Policy no.4 'Sustainable Development in Semi-Arid Sub-Saharan Africa'.

Brouwers, J.H.A.M., 1993. Rural people's response to soil fertility decline: The Adja case (Benin). PhD-thesis Wageningen University, P.O.Box 9109, 6700 HB Wageningen, The Netherlands. 157 p.

Fresco, L.O. and S.B. Kroonenberg, 1992. Time and spatial scales in ecological sustainability. Land Use Policy: 155-168.

Fresco, L.O., L. Stroosnijder, J. Bouma and H. van Keulen (eds.), 1994. The Future of the Land; Mobilizing and integrating knowledge for land use options. John Wiley Ltd, 375 p.

Friedel, M.H., 1991. Range condition assessment and the concept of thresholds: a viewpoint. Journal of Range Management 44(5): 425-426.

Hudson, N. and R.J. Cheatle, 1993. Working with farmers for better land husbandry. Soil and Water Conservation Society, Ankeny, USA, 272 p.

Huibers, F.P. and L. Stroosnijder, 1992. Irrigation and Water Conservation as Complementary Technologies in the Semi-Arid Tropics. P. 257-274 in: Diemer, G. and J. Slabbers (eds.) Irrigators and Engineers, Thesis Publishers Amsterdam, The Netherlands.

Jones, M.J., 1993. Sustainable agriculture: an explanation of a concept. P. 30-47 in: Crop protection and sustainable agriculture (Ciba Foundation Symposium), John Wiley Ltd.

Kessler, J.J., 1994. The usefulness of the human carrying capacity concept in assessing ecological sustainability of land-use in semi-arid regions. Accepted in: Agriculture, Environment and Ecosystems.

Keulen. H. van and H. Breman, 1990. Agricultural development in the West African Sahelian region: a cure against land hunger? Agriculture, Ecosystems and Environment, 32: 177-197.

Leeuw, P.N. de and J. Tothill, 1990. The Concept of Rangeland Carrying Capacity in Sub-Saharan Africa, Myth or Reality. Pastoral Development Network Paper 29b, Overseas Development Institute, London.

Pieri, Ch., 1989. Fertilité des terres de savannas; bilan de trente ans de recherche et de développement agricoles au sud du Sahara. Ministère de la Coopération et CIRAD-IRAT, France, 444 p.

Pol, F. van der, 1992. Soil Mining, an unseen contributor to farm income in Southern Mali. Bulletin no. 325, Royal Tropical Institute (KIT), Amsterdam.

Prins, H.H.T., 1992. The pastoral road to extinction: Competition between wildlife and traditional pastoralism in East Africa. Environmental Conservation, 19(2): 117-123.

Rabbinge, R., 1993. The ecological background of food production. P. 2-29 in: Crop protection and sustainable agriculture (Ciba Foundation Symposium), John Wiley Ltd.

Reijntjes, C., B. Haverkort and A. Waters-Bayer, 1992. Farming for the Future: An Introduction to Low-External-Input and Sustainable Agriculture. ETC/ILEIA and MacMillan Press Ltd, London UK.

Ridder, N. de, L. Stroosnijder, A.M. Cissé and H. van Keulen, 1982. Productivity of Sahelian Rangelands: A study of the soils, the vegetation and the exploitation of the natural resource. Volume I, Theory. Wageningen University, Nieuwe Kanaal 11, 6709 PA Wageningen, The Netherlands, 231 p.

Rietkerk, M., 1994. Hysteresis and thresholds in Sahelian Rangeland Development, PhD-research proposal, Wageningen University, 6709 PA, Wageningen, The Netherlands.

Sivakumar, M.V.K., 1992. Climate change and implications for agriculture in Niger. Climatic Change 20: 297-312.

Stroosnijder, L., 1992. Desertification in Sahelian Africa. The EEC Courier, 133: 36-39.

Stroosnijder, L., 1994. Modelling the effect of grazing on the soil water balance and the primary production in the Sahel. Submitted to Modelling of Geo-Biosphere Processes.

Stroosnijder, L., 1982. La pédologie du Sahel et du terrain d'étude. P. 52-71 dans: F.W.T. Penning de Vries & M.A. Djitèye (eds.), La productivité des pâturages Sahéliens. Agric. Res. Rep. 918, PUDOC, Wageningen.

Stroosnijder, L. and W.B. Hoogmoed, 1984. Crust formation on sandy soils in the Sahel; II: Tillage and its effects on the water balance. Soil & Tillage Research, 4:321-337.

Stroosnijder, L. and T. van Rheenen, 1993. Making Farming Systems Analysis a more objective and quantitative research tool. P. 341-353 in: F.W.T. Penning de Vries (ed.), Systems Approaches for Agricultural Development, Kluwer Academic Publishers.

Stroosnijder, L. and W.B. Hoogmoed, 1993. Sustainable Land Use in the Tropics; Management of Natural Resources in the Sahel, Interdisciplinary Research Programme 1994-1998. Wageningen University, Nieuwe Kanaal 11, 6709 PA Wageningen, The Netherlands.

Stroosnijder, L., W.B. Hoogmoed and J.A.A. Berkhout, 1994. Modelling effects of water conservation tillage in the semi arid tropics. International symposium (1991) on 'Gestion agroclimatologiques des précipitations, une voie de réduction du gap technologie de l'agriculture tropical africaine. Bamako, Republique du Mali. Sécheresse (in press).

Stroosnijder, L., L.O. Fresco and S.W. Duiker, 1994a. The Future of the Land. Submitted to: Land Use Policy.

Stroosnijder, L., S. Efdé, T. van Rheenen and L. Agustina, 1994b. QFSA: A new method for farm level planning. Proceedings International Conference 'Future of the Land', John Wiley Ltd.

Thom, R., 1978. Structural stability and morphogenesis: an outline of a general theory of models. (4th pr.). Benjamin Cummings, Reading (Mass.), 348 p.

Tiffen, M., M. Mortimore and F. Gichuki, 1993. More people, less erosion; Environmental Recovery in Kenya. John Wiley Ltd, 311 p.

World Bank, 1989. Sub-Saharan Africa, from Crisis to Sustainable Growth: a Long Term Perspective Study, IBRD/World Bank, Washington, USA.

Population, Environnement et Economie au Sahel

Serge Snrech

Club du Sahel, OCDE, France

Le Sahel connaît des changements rapides

La population du Sahel croît rapidement: d'une population d'environ 11 millions de personnes en 1930, elle est passée à 21 millions en 1960 et avoisine aujourd'hui les 45 millions.

Ce rapide accroissement démographique s'est accompagné, dès le début du siècle dans certaines régions, mais de façon beaucoup plus systématique dans les trente dernières années, d'une monétarisation croissante de l'économie.

Au coeur de l'économie de marché, à l'interface entre le Sahel et le reste du monde, les villes sahéliennes se sont fortement développées: le taux d'urbanisation est en effet passé de 4% en 1930 à 10% en 1960 et 26% en 1990. Le nombre d'urbains a ainsi été multiplié par 5 entre 1960 et 1990, passant de 2 à 10 millions.

La croissance de la population

La croissance démographique est un phénomène ayant une grande inertie. Quelles que soient les mesures d'encouragement du planning familial qui seront prises (et il est urgent qu'elles soient prises), le doublement de la population sahélienne dans les trente prochaines années est déjà inscrit dans la structure de la pyramide des âges (cf fig. 1).

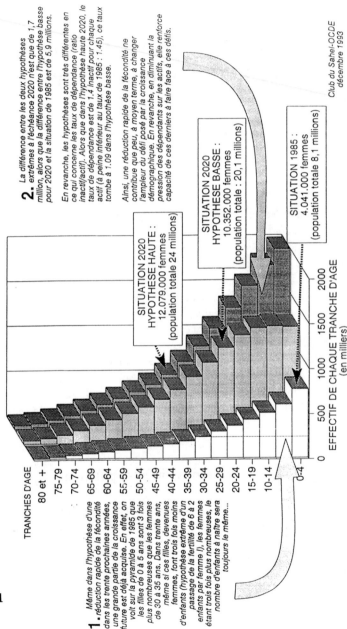

L'INERTIE DES PHENOMENES DEMOGRAPHIQUES
EXEMPLE DE LA PYRAMIDE DES AGES DES FEMMES AU MALI :
SITUATION 1985, HYPOTHESE BASSE ET HYPOTHESE HAUTE POUR 2020

Fig. 1

1. Même dans l'hypothèse d'une réduction rapide de la fécondité dans les trente prochaines années, une grande partie de la croissance future est déjà acquise. En effet, on voit sur la pyramide de 1985 que les filles de 0 à 5 ans sont 3 fois plus nombreuses que les femmes de 30 à 35 ans. Dans trente ans, même si ces filles, devenues femmes, font trois fois moins d'enfants (hypothèse extrême d'un passage de la fertilité de 6 à 2 enfants par femme !), les femmes étant trois fois plus nombreuses, le nombre d'enfants à naître sera toujours le même...

2. La différence entre les deux hypothèses extrêmes à l'échéance 2020 n'est que de 1,7 million, alors que la différence entre l'hypothèse basse pour 2020 et la situation de 1985 est de 5,9 millions.

En revanche, les hypothèses sont très différentes en ce qui concerne les taux de dépendance (ratio inactif/actif). Alors que dans l'hypothèse haute 2020, le taux de dépendance est de 1,4 inactif pour chaque actif (à peine inférieur au taux de 1985 : 1,45), ce taux tombe à 1,09 dans l'hypothèse basse.

Ainsi, une réduction rapide de la fécondité ne contribue que peu, à moyen terme, à changer l'ampleur du défi posé par la croissance démographique. En revanche, en diminuant la pression des dépendants sur les actifs, elle renforce capacité de ces derniers à faire face à ces défis.

Club du Sahel-OCDE
décembre 1993

En effet, du fait d'une forte fécondité (environ 7 enfants par femme, soit 3,5 filles par femme), les petites filles qui feront des enfants dans 30 ans seront beaucoup plus nombreuses que leurs mères. Même si le nombre d'enfants par femme est fortement réduit à ce moment, l'augmentation du nombre de mères empêchera un ralentissement rapide de la croissance démographique.

Taux de fécondité dans quelques pays					
PAYS	TAUX	PAYS	TAUX	PAYS	TAUX
Burkina Faso	7,2	Kenya	6,5	Sri-Lanka	2,9
Sénégal	6,3	Zaïre	6,1	Chili	2,7
Niger	7,1	Maroc	4,2	Chine	2,2
Mali	7,3	El Salvador	4,6	Etats-Unis	2,0
		Bangladesh	4,9	Danemark	1,6

En fait, dans les années passées, seule l'émigration a permis de réguler à court terme l'accroissement de population dans les pays sahéliens.

L'émigration vers les pays côtiers ou le reste du monde, a permis de limiter la croissance totale des effectifs dans les pays sahéliens. Mais, après plusieurs décennies de migrations intensives, les pays côtiers (et plus encore, mais pour d'autres raisons, les pays industrialisés) commencent à saturer et se montrent de moins en moins accueillants pour les populations sahéliennes: on peut donc de moins en moins compter sur ces pays pour absorber le trop plein de croissance démographique des pays sahéliens.

L'émigration a également contribué à redessiner la carte de la population à l'intérieur des pays sahéliens. Toutes les zones n'ont pas connu les mêmes taux de croissance réelle. Les zones les plus enclavées et/ou les moins bien dotées en ressources naturelles ont connu un exode plus important. Certaines zones particulièrement difficiles (en Mauritanie notamment), ont même connu une baisse de population. Mais cela reste exceptionnel.

Pour un taux d'accroissement naturel de l'ordre de 3% par an, le monde rural sahélien a connu un accroissement réel de 2% (le reste contribuant à l'exode rural ou l'émigration internationale). Mais cet accroissement moyen cache de fortes disparités. Les taux de croissance constatés dans les régions varient pour l'essentiel entre 1% et 4% par an. Durant les trente dernières années, ce sont les régions les plus au sud (zones cotonnières) et les régions proches des grandes villes qui ont connu la plus forte croissance.

Population totale des pays du CILSS: évolution 1930-1990 et projection 2020

	pop.totale	pop. urbaine	pop.rurale	tx d'urban.
1930	11.5	0.4	11.1	3.6
1960	21.6	2.1	19.5	9.9
1970	26.7	3.9	22.8	14.7
1980	33.1	7	26.1	21.2
1990	41.6	10.7	30.9	25.8
2020	79.1	43.3	35.8	54.8

Les systèmes traditionnels

L'ensemble des systèmes de production anciens était basé sur le même principe: la valorisation du renouvellement marginal du stock de ressources naturelles préexistantes. Tout écosystème fixe en effet, chaque année, un ensemble d'éléments minéraux indispensables à la production biologique (et en particulier agricole), comme un compte en banque qui produirait des intérêts à partir d'un capital initial.

Les systèmes de production anciens sont basés sur la valorisation de cette production marginale. Le système le plus simple, pour cela, est la culture itinérante: on cultive pendant 3 à 5 ans (et pendant ce temps là on consomme une bonne partie du capital disponible), puis on laisse le système se reconstituer, pendant 15 à 20 ans, voire plus. L'intégration de l'agriculture et de l'élevage constitue un autre de ces systèmes: les animaux prélèvent le capital naturel disponible dans la brousse et le concentrent, par leurs déjections, dans les parcelles cultivées: on a une péréquation dans l'espace de la fertilité.

Tous ces systèmes ont évidemment pour limite physique la vitesse de renouvellement naturel du stock de ressources naturelles. Certaines techniques peuvent être utilisées pour maintenir cette limite à un niveau élevé: elles visent soit à limiter les pertes d'éléments sous culture (il s'agit notamment de toutes les mesures de contrôle de l'érosion, et de tous les mécanismes de restitution des déchets végétaux), soit la stimulation de processus naturels (jachères plantées de légumineuses, agro-foresterie...).

Mais, même avec ces techniques (qui ont pour la plupart déjà été mises en oeuvre par les populations dans divers systèmes de production lorsque cela s'est avéré nécessaire), lorsque la ponction s'accroît, on finit par butter sur la capacité de renouvellement spontané des ressources naturelles. En effet, les mécanismes à l'oeuvre dans ce renouvellement sont particulièrement faibles au Sahel: les mécanismes physico-chimiques (dégradation des roches sous-jacentes) buttent sur la pauvreté des roches (qui sont pour la plupart des grès anciens); les mécanismes biologiques buttent sur l'extrême sécheresse qui sévit pendant plus de la moitié de l'année sur la région, empêchant la plupart des mécanismes de s'instaurer durablement.

La ponction sur le système a fortement augmenté

Les systèmes de production anciens fonctionnaient bien dans le passé. Mais la région était peu peuplée, le contrôle social sur les individus était fort, les besoins de consommation peu développés et les alternatives à l'agriculture quasiment inexistantes. Tous ces paramètres sont en évolution profonde et rapide: notamment, la population a fortement augmenté (cf fig. 2), et la consommation s'est fortement extravertie: les ruraux consomment de plus en plus de biens qu'ils ne produisent pas et doivent donc acheter.

Pour faire face à cette double croissance des besoins, les ruraux ont fortement augmenté leur production agricole, pastorale et forestière. Cela s'est traduit dans beaucoup de régions par une forte ponction sur les ressources naturelles, et une dégradation visible de l'environnement.

EVOLUTION DE LA DENSITE DE POPULATION RURALE

EN AFRIQUE DE L'OUEST ENTRE 1960 ET 1990

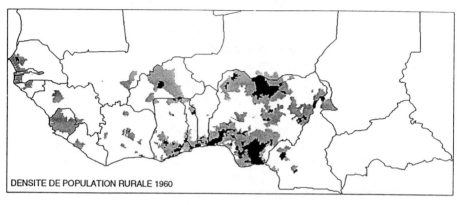

DENSITE DE POPULATION RURALE 1960

LEGENDE

☐ densité inférieure à 20 hab/km2

▦ densité comprise entre 20 et 60 hab/km2

■ densité supérieure à 60 hab/km2

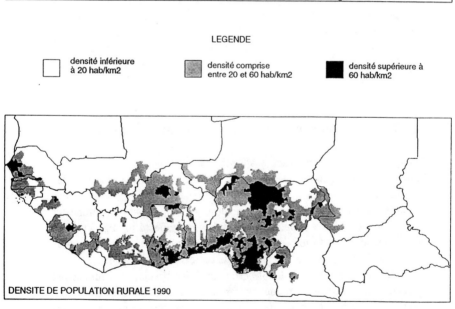

DENSITE DE POPULATION RURALE 1990

Fig. 2

Source : ETUDE BAD/OCDE WALTPS, 1994

Au-delà du constat de cette dégradation visible, le débat est ouvert pour savoir si elle résulte d'une mauvaise gestion des ressources naturelles, ou d'une saturation des capacités de production biologique plus directement liée à la nature même des écosystèmes. Selon plusieurs experts (Graff & Breman 1993; FAO 1986) beaucoup de régions consomment désormais plus que les écosystèmes ne sont capables de renouveler *dans l'absolu*. Il est probable que plusieurs régions ont effet atteint cette capacité de charge absolue, ou sont sur le point de la dépasser.

Difficile de maintenir les systèmes très intensifs en main d'oeuvre

A la saturation progressive de la capacité de charge écologique s'ajoutent les transformations du contexte économique et social dans lequel s'insèrent les agricultures. Les systèmes anciens étaient performants écologiquement, mais c'était au prix d'une forte pression de la société sur l'individu. Aujourd'hui, les systèmes sociaux se sont fortement assouplis. Parce que la mobilité est très élevée au Sahel (les ruraux sahéliens voyagent souvent pendant la saison sèche, et connaissent tous "la ville"), parce que l'information circule beaucoup plus, et beaucoup plus vite (du fait du transistor et, de plus en plus, de la télévision), les Sahéliens, en particulier les jeunes, ne sont plus prêts à fournir d'importants efforts sans une rémunération acceptable.

Or, une rémunération acceptable aujourd'hui, est une rémunération qui s'établit sur un marché du travail de plus en plus fluide. Il existe certes un "coût" pour quitter le village et, dans le cas de communautés très soudées, ce "coût social" du départ peut pousser les villageois à accepter des rémunérations notablement plus faibles qu'en ville. Il est vrai également que tout le monde ne dispose pas d'un réseau d'accueil lui permettant de s'installer facilement en ville ou dans des zones où il existe encore des terres vierges. Mais des signes nombreux et convergents laissent penser que l'agriculture subit de plus en plus la concurrence directe des autres possibilités de travail ou d'investissement. Ainsi, dans certains villages du Burkina Faso, les vieux se plaignent du fait que les jeunes exigent désormais d'être payés pour réaliser les travaux d'intérêt collectif, ce qui aurait été inimaginable quelques années plus tôt.

L'agriculture sahélienne à un tournant

L'agriculture sahélienne ne peut plus vivre durablement sans investissement massif, en main d'oeuvre et/ou en intrants. Par ailleurs, le contrôle social sur la main d'oeuvre n'est plus suffisant pour lui imposer des tâches lourdes qui seraient très fortement sous-rémunérées.

L'agriculture sahélienne arrive donc à un seuil où elle doit passer de la gestion des ressources naturelles à un système plus intensif de type input/output (un système où les produits exportés doivent être compensés par des apports équivalents d'éléments minéraux (donc un système plus intensif en capital)) et en même temps, elle doit se transformer en activité économique pleinement concurrentielle, au risque sinon de se voir condamner à la marginalisation.

Cette double mutation est évidemment difficile. Elle ne se fera que par les paysans (le rôle des aides et des Etats dans cette mutation est "marginal"), mais elle ne se fera certainement pas partout, ni avec tous les paysans.

En effet, pour entrer dans une agriculture intensive et durable, les paysans sont confrontés à une série d'arbitrages:

-    avec un revenu donné, le paysan doit d'abord arbitrer entre ce qu'il va investir dans des activités économiques, et ce qu'il va consommer (consommation qui correspond parfois à un investissement "social");

-    lorsqu'il investit dans les activités économiques, le paysan doit ensuite choisir entre investir dans l'agriculture ou diversifier ses activités (investissement dans le commerce, les services, l'immobilier...);

L'ETAT ET LES PAYSANS DANS LA GESTION DES RESSOURCES NATURELLES :
REPRESENTATION SCHEMATIQUE DES ROLES ET INTERACTIONS

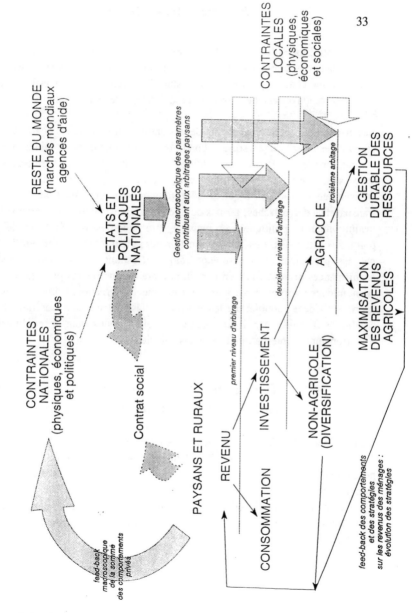

S. Snrech
Club du Sahel-OCDE
nov. 1993

Fig. 3

-       enfin, lorsqu'il investit dans l'agriculture, il a le choix entre investir dans des pratiques qui maximisent son profit à court terme (achat d'une charrue pour doubler la superficie cultivée) ou investir dans le renouvellement à long terme de la fertilité (engrais etc.).

Les critères qui interviennent lors de ces choix sont nombreux: conditions climatiques et de sols, conditions économiques, conditions institutionnelles... Parmi les critères qui contribuent aux arbitrages successifs, certains sont purement locaux (conditions physiques, environnement social), d'autres sont plus macroscopiques (fiabilité et transparence des marchés, termes de l'échange villes/campagnes, contexte institutionnel et politique), et peuvent être modifiés par le biais des politiques nationales. Les politiques nationales, à leur tour, sont influencées par les performances macro-économiques (résultat de la somme des performances privées), par un certain nombre de contraintes politiques et économiques internes et par leur environnement international. Le rôle des agences d'aide est double: par leurs projets, elles contribuent à modifier les conditions locales; par leurs politiques, elles contribuent à modifier l'environnement international des pays sahéliens.

Tous les territoires sahéliens ne sont pas dans la même situation

Hétérogénéité des densités de peuplement, hétérogénéité des conditions écologiques, hétérogénéité de la relation au marché: les problèmes se posent dans des termes très différents selon les zones des pays sahéliens où l'on se trouve.

Il est évidemment difficile et toujours arbitraire de distinguer des zones de potentiel économique. On propose néanmoins ici deux classifications possibles:

A: on peut distinguer trois zones d'aptitude agro-climatiques contrastées:

- *les zones à plus de 700 mm de précipitations où est pratiquée la culture du coton*: ce sont des zones relativement riches, disposant jusqu'à présent de débouchés garantis et de prix rémunérateurs. Ces zones disposent de fortes incitations à la production; elles se caractérisent pour l'instant par un développement très extensif, mais elles peuvent, sous certaines conditions, accueillir une agriculture intensive, durable et néanmoins rentable (Delgado 1991).

- *les zones entre 700 et 400 mm* : elles sont caractérisées par une économie de subsistance, essentiellement basée sur les céréales et l'élevage; ce sont des zones fortement peuplées (nord du bassin arachidier au Sénégal, plateau Mossi au Burkina, pays Haoussa au Niger), qui ne produisent pas de produits de rente à destination du marché mondial (à l'exception d'un peu d'arachide), mais disposent néanmoins d'un réel potentiel de production pour le vivrier. Dans ces zones, c'est la capacité d'obtenir un marché stable et suffisamment rémunérateur pour les produits vivriers qui conditionnera à l'avenir l'émergence d'une agriculture durable.

- *les zones en dessous de 400 mm*: très faible potentiel pour l'économie de marché (faible marché, faible réponse à l'investissement...), à part pour l'élevage extensif. La population de ces zones, faiblement solvable, essaie de produire pour maximiser la part d'autoconsommation. Cela amène les populations à produire des céréales, contre l'avantage comparatif de la région qui voudrait qu'on y fasse plutôt de l'élevage. Ce sont des zones de forte émigration, dont on peut sérieusement mettre en doute l'avenir agricole.

B. On peut également zoner l'espace en fonction de l'influence qu'y exerce le marché:

-       la capacité des exploitations agricoles de faire face aux défis
        économiques évoqués ci-dessous est en effet la combinaison de deux
        facteurs: les conditions physiques de la production, croisées avec
        les opportunités économiques de production. On a vu que le coton
        constitue la base économique des régions les plus méridionales des
        pays sahéliens, mais que les autres zones dépendaient avant tout de
        la demande en produits vivriers. On a donc modélisé la demande en
        produits vivriers des villes ouest-africaines, et la façon dont cette
        demande se transmet dans l'espace (cf fig. 4).

MODELISATION DE LA RELATION AUX MARCHES URBAINS

DES ESPACES RURAUX OUEST-AFRICAINS EN 1990

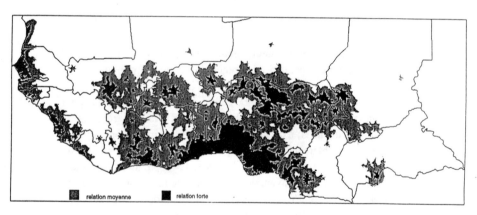

relation moyenne   relation forte

SOURCE : WALTPS BAD/OCDE, 1994   **Fig. 4**

Les test effectués sur le modèle montrent que, à l'exception des zones trop
sèches, on a une bonne corrélation entre le degré de connexion au marché
urbain et la productivité per capita, ce qui montre l'importance des marché
dans le processus de production dès aujourd'hui.

Priorités d'affectation des fonds disponibles pour l'investissement

Deux rôles principaux existent pour l'aide: une aide "au développement", qui aide les populations à effectuer les nécessaires transformations dans leurs modes de vie et de production; une aide "sociale", qui maintient les population au-dessus d'un niveau minimum (lui même variable) mais sans espoir d'amélioration notable de leur capacité à subvenir à leur besoins.

Il est difficile de définir le juste équilibre entre ces deux types d'aide. Les analystes qui comparent le processus de développement en Asie et en Afrique (notamment J. Mellor 1991 et U. Lele 1989) soulignent toutefois l'importance de l'aide sociale dans les politiques d'aide à l'Afrique, qui s'est parfois faite au détriment de la croissance des pôles les plus dynamiques qui aurait également fini par rejaillir sur les zones pauvres.

L'aide doit donc choisir les thèmes sur lesquels elle veut intervenir, mais aussi les lieux sur lesquels elle souhaite intervenir. Là encore, le dilemme est difficile à résoudre. La répartition des populations sahéliennes est en effet très hétérogène: d'un côté, la moitié la plus dense de la population rurale est concentrée dans moins de 10% de l'espace: elle est accessible à peu de coût, mais souvent moins nécessiteuse que les populations enclavées. A l'autre extrême, un quart de la population se répartit sur plus de 75% du territoire des pays sahéliens: il s'agit de populations qui auraient probablement plus besoin d'aide, mais dont le coût d'approche individuel est, du fait de leur faible densité, extrêmement élevé.

La question du choix de l'affectation de l'aide, dans un contexte où les besoins sont immenses et les disponibilités limitées est évidemment une question très délicate: elle devrait recevoir une haute priorité des agences de coopération.

L'efficacité maximale de l'aide

En tout état de cause, c'est probablement dans les situations de transition que l'efficacité des interventions extérieures peut être la plus importante. On a vu en effet que la plupart des agricultures traditionnelles étaient extensives. La modernisation impose le passage à une certaine intensification.

Or, tant que les ressources naturelles sont disponibles en quantité, il est généralement beaucoup plus rationnel, économiquement, d'augmenter les superficies cultivées (et donc la collecte de capital naturel) que d'intensifier (ce qui oblige à augmenter les apports en travail et capital). Cependant, lorsque les ressources sont trop fortement dégradées, il devient difficile de valoriser les intrants de base de l'intensification (voir, par exemple, la très faible réactivité aux engrais des sols trop pauvres en matière organique).

La transition n'est probablement possible que dans un court laps de temps, pendant lequel la pression physique et économique est suffisante pour pousser à modifier les pratiques anciennes, tandis que les sols ne sont pas encore trop dégradés et permettent de maintenir une certain rentabilité des exploitations agricoles dans leur nouveau fonctionnement. Il s'agit d'une porte étroite, que l'on peut facilement rater.

Par exemple, Uma Lele (1989) a montré, en comparant plusieurs expériences de régions cotonnières, que le coton en Afrique anglophone (Nigéria, Ghana), peu intensif (peu d'input, peu d'output), avait un meilleur avantage comparatif que celui des pays francophones beaucoup plus intensif en produit importés. Cependant, lorsque l'espace se densifie et que les cultures vivrières deviennent, de ce fait, plus avantageuses, les régions cotonnières anglophones abandonnent le coton pour les vivriers. Cette monoculture vivrière, rentable à court terme, interdit en fait une intensification durable des systèmes de production. Au contraire, les zones cultivant le coton intensivement ont un potentiel de réponse bien plus important en matière céréalière; l'intensification sur coton, coûteuse en termes macro-économiques, pourrait donc être perçue comme un investissement permettant des transformations structurelles positives.

Quelques questions

Le système d'analyse dont on a esquissé les contours ci-dessus est déjà complexe. Pourtant, il devrait encore être complété par une analyse approfondie de plusieurs questions. Parmi les plus urgentes, on peut citer:

- *Quel sera l'effet des migrations de retour?* Ces migrations sont importantes du fait de la conjoncture. Les migrants qui reviennent aujourd'hui dans leurs villages, avec des connaissances et des capitaux importants, auront-ils un effet notable sur les évolutions des systèmes de production et, au-delà, sur la perception de l'activité paysanne?

- *Comment va évoluer la solvabilité des villes, et leur capacité à accueillir des migrants?* La "bonne santé" des villes est une inconnue majeure de l'avenir, et un facteur doublement important pour l'évolution de l'agriculture sahélienne: des villes en bonne santé économique auront une demande alimentaire importante et variée, fournissant autant d'opportunités économiques aux producteurs agricoles. Ces villes pourront accueillir le trop plein de population rurale qu'elle transformeront en client. En effet, si on peut imaginer des scénarios de développement durable de l'agriculture sahélienne, il est exclu qu'un scénario de développement agricole durable soit possible avec 75% de ruraux et seulement 25% d'urbains consommateurs...

Conclusion

Le Sahel traverse une phase cruciale de son histoire où il doit effectuer, dans un temps très bref et avec des conditions économiques difficiles, des mutations profondes de ses structures de production agricole. Ces mutations se font sous la contrainte des événements; leur coût social est immense, les besoins qu'elles engendrent également et leur issue reste incertaine. Toutes les régions et tous les paysans ne sont pas égaux face à ces mutations et, quelle que soit son issue, on sait que le processus sera fortement inégalitaire.

Dans ce contexte, les interventions extérieures doivent avant tout concevoir qu'elles ne représentent qu'une partie minime de l'effort total qui devra être fourni pour mener à bien cette transformation. Mais, tandis que la plupart des efforts "privés" des individus sont très largement prédéterminés par leur environnement, les agences d'aide disposent d'une très grande liberté d'affectation de leurs moyens.

La liberté dont disposent les agences d'aide, au sein d'un système extrêmement contraint par ailleurs, devrait les inciter à réfléchir à la meilleure utilisation qu'ils pourraient faire de leur argent pour catalyser le processus de transformation tout en le maintenant dans un domaine socialement acceptable. L'approche multidisciplinaire et systémique proposée laisse entrevoir que la voie de développement optimale de la région dans son ensemble est très loin de ce que l'on pourrait souhaiter atteindre dans chaque secteur ou dans chaque petite région. Elle montre aussi que l'aide a une efficacité maximale dans les zones en transition, et c'est peut-être dans ces zones qu'elle devrait concentrer ses actions.

Références

Delgado, Christopher, 1991. Voir l'article dans les actes du séminaire CIRAD/Club du Sahel sur l'avenir de l'agriculture sahélienne, Montpellier - septembre.

FAO, 1986. Agriculture africaine: les 25 prochaines années, Rome.

Graff, van der & Breman, 1993. Agricultural production: ecological limits and possibilities, CABO-DLO, Wageningen.

Lele, Uma et Steven W. Stone, 1989. Population Pressure, the Environment and Agricultural Intensification. Variations on the BOSERUP Hypothesis. Madia Discussion Paper (4), The World Bank.

Mellor, John W., 1991. Agricultural growth in Asia and Africa: the Population, Urbanization, Poverty Environmental Interaction, IFPRI.

Population in development and environment

Bertil Egerö
Programme on Population and Development
University of Lund, Sweden

PROP - A brief presentation

The Programme on Population and Development, PROP, was created in 1990 under an agreement between Swedish SIDA and the Department of Sociology, University of Lund. Its mandate is to strengthen the Swedish human resource base in the social sciences in the subject field of population/development relations. PROP is a networking organisation, and a center of documentation and publication. It serves SIDA, the government, and other institutions in Sweden.

PROP was created because of the gradual realization within SIDA and other quarters that population is not just the delivery of contraceptives. Demographic dynamics is part of social change - an effect as well as a cause of change in society. Its relation to development is complex and manifold. A better understanding of the relation of population to development is needed in the development cooperation, and this requires contributions from the social sciences. However, the appropriate competence is difficult to find - virtually all Swedish demographers are specialized on industrialized societies.

The rationale for PROP - the current lack of skills in social science - also impedes its expansion. No Swedish university offers the institutional capacity required for cooperation with counterparts in the South in research and higher education. No regular courses are offered in demography or population studies with an emphasis on developing countries. As a result,

PROP has to support Swedish students in finding suitable courses elsewhere, primarily in Europe. Indeed, we want to recommend Swedish students interested in development issues in the South to apply for admission at universities in the South. This requires an identification of universities for students used to high standards of educational and library services, housing etc., and a solution to the financial means required.

As implied by this brief presentation of the Programme, it is still primarily inward-looking, directing its activities to the Swedish (and Scandinavian) needs, not least those of agencies involved in development cooperation. One important activity of the Programme consists of improving the links with scientific work done in the South, and feeding scientific knowledge into the process of development cooperation.

SIDA and population-related development issues

Like most government agencies for development aid, SIDA has for many years regarded fertility and birth control as the central issue in population assistance. Modern contraceptive services, supplemented by information and motivation inputs, have been the main operational definition of population assistance. In recent years, partly in response to a perceived slowdown in fertility decline in the South, aid policies have been widened to incorporate various factors to reproductive behaviour. The factors include women's education and infant and child mortality. Currently, SIDA explores the importance of factors subsumed under headings such as 'reproductive and sexual health', 'empowerment', and 'human and reproductive rights'. To improve human conditions in these respects are development goals in their own right. In addition, the expectation is that such improvements will lead people to chose smaller families.

Current Swedish activities in the field of development cooperation strategy are spurred by the discussions preceding the UN International Conference on Population and Development to take place in Cairo in September this year. To the extent that contributions by SIDA and likeminded agencies and organizations will steer Conference deliberations on fertility trends into areas covered by the above headings, the conclusions and recommendations from the Conference are likely to cover areas well beyond the traditional field of family planning.

In this respect, the SIDA initiative is important and welcome. At the same time, the approach chosen addresses only one sector of the wide field of population/development relations, namely fertility change. Further, reproductive change is a longterm process, often inter-generational rather than simultaneous for older and younger generations. Sustained reproductive change, like real progress in public health, empowerment and rights, is part of a wider social and economic transition affecting social institutions such as the family, including the role and value of children. Thus, with full appreciation of the value of reforms to improve the general welfare of women and men, the very conditions under which such reforms are carried out remain to be addressed.

The relation between population and environment

Not long after the theory of the demographic transition had been generally accepted as a valid framework for understanding demographic dynamics under modern development, a new debate raged about population/development relations. This debate, rather than accepting demographic transition as part of modernization, centred on the conditions for poor dependent economies to take off in the direction of industrialization and social modernization, and the negative influence of high rates of population increase. Ansley Coale's and Edgar Hoover's study of India published in 1958 was a major contribution to the debate. Its conclusions supported arguments for separate government actions to bring birth rates down in favour of economic growth. Over time, the limitations to the Coale-Hoover analysis were exposed both in scientific research by, among others Simon Kuznets (1967), and through the historical fact of rapid 'take-off' of South East Asian economies in the midst of high rates of population growth.

The population and environment debate took longer to get underway. Rachel Carson's early alarm on environmental deterioration, Silent Spring (1963), was mainly about the effects of industrialization and wasteful consumption. Paul Ehrlich (1968) and Georg Borgström (1962) sounded early warnings about the threat of over-population in poor countries in relation to natural resources and food production. Their warnings were, however, temporarily contradicted by the economic advances in parts of the South, and by the success of innovations such as the Green Revolution.

In the history of Europe, vast environmental changes have taken place without much attention to the population aspects of such changes. The Greek islands long ago lost their forest cover and fresh-water sources. Virtually all the original European forests have disappeared under the pressure of human expansion and agricultural development. The Dutch have in recent centuries managed to enlarge their territory in Europe through the use of millions of trunks from tropical forests in their East Asian colonies. Swedish iron manufacture introduced the exploitation of vast tracts of forests in northern Sweden in its hunt for ever larger volumes of fuel wood. The paper and wood industry has since profoundly changed the quality of the forests and the character of the landscape.

Such environmental changes were related to waves of economic development carried by increasing numbers of human beings, an increase which itself was made possible by the economic advances. In other words, the environmental changes in Europe were brought about by an interaction between economic growth, demographic change, and social transformation. The population increase continued into the present century, where it is gradually approaching zero-level. The greater part of current environmental hazards is therefore linked to very high and still increasing per capita levels of wasteful and polluting nature consumption.

In discussions of the environmental change taking place in the South, economic development and social change generally do not get much attention. Rather, most writers see environmental problems through their 'demographic glasses', i.e. in relation to population numbers. In focus of this population and environment debate are the problems of high rates of population increase combined with rural poverty-driven overexploitation of scarce natural resources. Perhaps more than many other contemporary issues, this choice of perspective reflects the continuous North-South divide. It diverts attention from the badly needed discussion of what constitutes sustainable development in place of the obviously unsustainable

Western consumer society. On a local level, it detracts attention from the environmental impacts of large-scale exploitation of natural resources to the benefit of consumer societies growing in the midst of poverty. For the North, it provides a convenient shield against any demands that the rich states should reduce their populations, let alone put a halt to further increase, in response to heavy *per capita* pressures on the local commons as well as the global.

The North/South bias in the prevailing definition of population/environment relations may - and indeed does - provoke reactions which tend to favour the opposite position; that increasing human numbers bear no direct relation to natural resource depletion and other environmental changes. To me this is to replace one simplification with another. Demographic dynamics, a result of other changes in society, is itself a societal force in the conglomerate of forces acting for or against change. This does not mean, however, that a generalised expression can be formulated of the effects of this force, whether economic or environmental. As summarized in a recent review by Norman Myers, "the 'linkages calculus' is attended by uncertainty of many sorts." (Myers 1993:3)

Many if not most of the contributions to the population and environment debate appear to be motivated not so much by pure scientific interest as by policy issues. Often this agenda is implicit in the argument, a by-line conclusion or one the reader is led to draw. In the case of the Norman Myers text just referred to, the agenda is made explicit. Despite his initial caution on the character of the population/environment linkages, Myers is prepared to "offer the generalized conclusion that *population growth plays a prominent and probably predominant part in environmental problems*" (Ibid.:5; italics author's), where he includes as an example the ozone layer changes and the greenhouse effect. This leads Myers not unexpectedly to an identification of measures to reduce population growth as the first priority in stemming environmental deterioration.

In a local setting, specified in time and space, a more or less direct relation between changes in human numbers and environment change may be observed. But local settings are rarely isolated and cut off from external influence. External factors tend to affect the relation in significant ways, making population an intermediate variable, or the last in a series of causal connections bringing about the observed relation. Paul Harrison gives an illuminating discussion of land expropriation and marginalization of the poor:

> "Yet even where expropriation is the main driving force, the impact
> of population growth is powerful. For it is through population that
> inequality and expropriation work their impact on the environment.
> They confine the oppressed to a smaller area, and artificially boost
> population density. Natural population growth goes on to increase that
> density, and worsens the problem." (Harrison 1992:131)

Harrison's description of the process is correct. Yet, his formulation gives the impression of an underlying message, namely that population is the problem to be tackled if erosion of marginal lands should be avoided. Obviously, if there are fewer oppressed, then the environmental impact of their plight is correspondingly lighter. Still, there appear to exist more direct ways than birth control among the poor, for instance land reform, which could be expected to give more rapid results and at the same time influence reproduction through raising the living standards of the oppressed themselves.

An approach which holds out anti-natalist policies as the answer to environmental problems, whether in poor or in more resource-rich societies, has two fallacies. They can be formulated as follows:

-   First, in societies with high rates of population increase, even the most successful policy aimed at reducing fertility will achieve only a gradual reduction in the growth rates. The number of human beings will continue to increase for many years, unless halted by disasters.

-   Second, increasing numbers of people is an unavoidable ingredient in policies aiming at sustainable natural resource use. Whether and at what speed a 'fertility transition' to low levels will occur depends to a great extent on the social and economic conditions of current and future generations, i.e. on how well accessible natural resources are utilized in support of widespread and lasting welfare for steadily more numerous new generations.

Poverty and population/environment dimensions

How is this challenge to be met? Poverty is a characteristic not only of individuals. In today's Sub-Saharan Africa many states are broke and can neither serve the public, nor invest in the infrastructure necessary for growth. Under such circumstances, while many children may benefit the individual household, for the nation they constitute an additional constraint to those provided by external debt services and unequal terms of trade.

In nearly all countries in Sub-Saharan Africa, the majority of the inhabitants are smallholding farmers or agropastoralists. These have been the losers in development ever since colonial times. Often, the men have gone to work elsewhere for wages that did not compensate for the loss of their labour to the household economy. The additional burdens on the women have led them to economize their labour, favouring work with short-term returns at the cost of labour investment - in nutritionous crops, or in land conservation practices - that over time would pay off in healthier people and land.

In the absence of significant post-independence development in most rural areas, many men - and increasingly also women - have continued to migrate even though the chances of steady employment have shrunk with the economic crisis. The persistent lack of labour in rural households is one important factor among several which creates poverty through the efforts of the poor themselves. Another is the lack of access to technology, a third the lack of infrastructure to facilitate market contact and the selling of surplus agricultural production. Taken together, such constraints go some way to explain why the 'Boserupian' effect (1965) on production from population pressure is by and large absent from the smallholder sector in Sub-Saharan Africa. At the same time, the primary sector will for a majority of the population remain the key to economic development, and thereby also in the necessary transition to small families.

Rethinking development in favour of the smallholder economy should therefore be part of the agenda. For countries such as Tanzania, uneven distribution of population reflects historical and political circumstances at least as much as land use potential (Moore 1973). The main untapped resource for Tanzania lies in its land, which still in many areas has an insufficient population base for more efficient exploitation. As remarked by Ester Boserup herself (1985), it might make good economic sense to build roads and other infrastructure, clinics, and education services. This could help to establish an economically viable smallholder agriculture, which is capable to intensify land use in a sustainable way as well as maintain the social services. This comment certainly applies to relatively underpopulated areas with good agricultural potential.

Concluding comments

To conclude, there are two main areas of concern over environmental deterioration in the South. One, the least discussed and to my mind the most serious, is the destruction and pollution going with development as it is known in the North. This destruction comes from increasingly speculative exploitation of natural resources, pollution from manufacturing, consumption (including of private transport), and waste management. This threat is growing exponentially through the joint effects of population increase and increasing proportions of the population reaching such income levels that they can join the consumer society.

The other are the environmental effects of rural poverty carried by households which are constrained to insufficient natural resources and short of labour, technology, and tools to intensify exploitation in sustainable ways. Generally, the poor depend on their children and are happy to have more. High fertility steadily adds new members to the poor, increasing the competition for scarce resources and worsening the prospects for permanent improvements in the struggle against poverty.

Both of these threats are related to the increasing numbers of people. This increase reflects the conditions under which people live, and will change as living conditions change. The real challenge comes from the momentum of population growth, which will have to be accommodated in development strategies and management of available resources.

References

Borgström, G., 1962. Mat för miljarder. STs Förlag, Stockholm.

Boserup, E., 1985. Economic and Demographic Relations in sub-Saharan Africa. Pp. 383-397 in: Population and Development Review 11, no. 3.

Boserup, E., 1965. The Conditions of Agricultural Growth. The Economics of Agrarian Change under Population Pressure. George Allen & Unwin Ltd, London.

Carson, R., 1963. Silent Spring. H. Hamilton. London.

Coale, A.J. and Hoover, E.M., 1958. Population Growth and Economic Development in Low-Income Countries. Princeton Univ. Press.

Egerö, B., 1992. No Longer North and South - The New Challenges of Demographic-Economic Interrelations. - Pp. 85-98 in: Hansson, L.O. and Jungen, B. (Eds). Human Responsibility and Global Change. Proceedings from the International Conference in Göteborg 9-14 June, 1991. University of Göteborg, Section of Human Ecology.

Ehrlich, P., 1968. The Population Bomb. Ballantine, New York.

Harrison, P., 1992. The Third Revolution: Environment, Population and a Sustainable World. I.B.Tauris & Co Ltd.

Kuznets, S., 1967. Population and economic growth. Pp. 170-193 in: Proceedings of the American Philosophical Society 111, no. 3.

Moore, J.E., 1973. Population Distribution and Density. - Pp. 38-55 in: Egerö, B. and Henin, R.A. (Eds), The Population of Tanzania. An Analysis of the 1967 Population Census. BRALUP and Bureau of Statistics, Dar es Salaam.

Myers, N., 1993. Population, Environment, and Development. Reprint from: Environmental Conservation, Vol.20, Nr 3, Autumn 1993.

Population Issues in the Sahel

Peter Plesner
Centre for Development Research, Denmark

Introduction

Population growth in the Sahel has not been seen as a major constraint to development. The region has always been very scarcely populated, due to the climatic conditions. Nowadays, however, with one of the fastest growing populations in the world, the Sahel region is getting increasing attention in the population debate. Especially in relation to environmental issues population growth is seen as one of the major components in the degradation of the region.

The present population of a little more than 40 million is expected to more than double within the next 30 years. This will be the fastest growth ever seen in the history of the Sahel, and it will definitely influence the future development. The growth in rural areas can be expected to result in changes of the organization of both agricultural and non-agricultural production. Migration to towns and cities will play an even more important role in the future. The interaction between rural and urban areas will probably need to change in character and size due to the growing populations.

Understanding population growth, its impact on the region and searching for a future strategy implies searching for answers to the following questions:

-        What explains the growing population of the Sahel?

-        What are the region's prospects?

-         What plan would slow population growth?

These questions are discussed below. The truth is, however, that there is no simple solution to growing populations. Redundant, multidisciplinary approach is needed. At the risk of generalizing in a diversified region, the explanations outlined below are hoped to suggest further action.

Fertility in the Sahel

Fertility is determined by several factors and can not be explained simply. Factors such as poverty, education, the role and status of women, access to health care, family planning, and individual strategies for survival are known to affect population growth.

Sahel is among the poorest regions in the world (Table 1). With an average GNP per capita of about US$ 300, the region is poor in official terms. Even though there is no direct correlation between fertility rates and GNP within the region, poverty is of major importance to the high population growth.

Another striking feature is the extremely low figures for adult literacy. Only about one fifth of the women can read and write. People disseminating information about birth control need to take this into consideration as well.

The high illiteracy among women in the Sahel reinforces the role and status of women. With limited professional skills and low education, the status of a woman is likely to be determined by the number of children (sons) she can give birth to during her lifetime. The widespread polygamy in the region tends to further enhance the number of children as the major determinant for the status of women. Each wife often more or less compete with the other(s) in terms of the number of children they have.

The poverty of the region is also shown in the high infant mortality rate and low life expectancy. On average each women in the Sahel is likely to loose at least one baby before the child reaches its first year. The bad health conditions in the region indicate the limited access to health facilities. As family planning services are almost exclusively attached to

Selected Population and Social Indicators for the Sahel Countries

	POPULATION 1992/2025	AVERAGE GROWTH RATE (%) 1990-95	FERTILITY FATE PER WOMAN 1990-95	INFANT MORTALITY PER. 1000 1990-95	LIFE EXPECTANCY 1990-95	ADULT LITERACY M/F 1990	PER CENT URBAN 1992	GNP PER CAPITA (US$ 1990)
Burkina Faso	9.5/22.6	2.8	6.5	118	48	28/9	17	330
Chad	5.8/12.9	2.7	5.9	122	48	42/18	34	190
Mali	9.8/24.6	3.2	7.1	159	46	41/24	22	270
Niger	8.3/21.3	3.3	7.1	124	47	40/17	21	310
Senegal	7.7/17.1	2.7	6.1	80	49	52/25	41	710
Africa	681/1582	2.9	6.0	95	53	NA	33	NA
Developing countries	4254/7069	2,0	3,6	69	62	NA	35	NA

Tabel 1 (Source: UNFPA: State of World Population 1992)

health clinics in Africa, this indicates another area of concern in terms of reduction in the fertility levels.

Finally, it is evident that with the chance of loosing some of the children and the status determined by the number of boys, it is not unlikely, that 3-4 boys is seen as a minimum for security in old age. As girls in most of the Sahel move to the husband's family after marriage, only boys can take care of the parents when they become unable to work.

Prospects

During the past decade, the Sahel has been touched by drought. In a harsh environment this decline in rainfall has lead to overgrazing, soil degradation, and human suffering. In this situation, population growth has meant that more people have had to make a living on decreasing natural resources.

Carrying capacity is often defined as the number of people the land can support. This varies with the technological development. If population growth is followed by changing modes of production, natural resources could probably support the growing population for many years. Ester Boserup has previously shown how population growth in itself can determine such changes in agriculture. In most parts of the Sahel there are, however, few signs of such a development. Thus, the growing population combined with drought has degraded the land.

Migration from rural areas has been one of the consequences of increased population growth and land degradation. Towns and cities have grown with a pace that urban planning cannot cope with. The result has been the enlargement of slums and poverty. With the food shortage in many rural areas, food import will continue to play an important role in the future.

Population policies

Given its vastness, the Sahel is a geographic rather than cultural entity. As population policies interfere with people's right to have children, cultural differences need to be taken into consideration. Popular participation is essential for the success of population programmes. Because of the cultural differences, as such, it is unlikely that a general strategy could be adopted in the Sahel. There is, however, an urgent need for action. We need to consider the following about birth-control policies:

-   *Human rights*: No population program should force people to limit the number of children. Evidence from several Asian countries has shown that forced sterilization or promotion of civil servants according to the contraceptive prevalence in their area of responsibility, tends to have a negative impact. The question of human reproduction remains one of the most basic human rights. It should, therefore, remain the choice of the individual family.

-   *Information, education, and communication*: The education and communication should include also the younger age groups. Information about sex needs to address taboos. Even though such information programmes might interfere with religious practices, it is often necessary that birth-control programmes take the lead in the openness of the questions.

-   *Availability of contraceptives*: Availability of contraceptives do not necessarily do the job. First, the decision about whether to have a child needs to be made. The question of contraception comes second, which seems to be forgotten among some donors. It is, however, no secret that modern contraceptives are more efficient than traditional. In addition, traditional contraceptive methods have lost a great deal of their previous prevalence. But modern contraceptives are almost all associated with health risks. Therefore, the distribution of modern contraceptives has to take place in collaboration with the health care system. The availability of contraceptives needs to be as varied as possible and users needs to be informed on how the contraceptives work and which adverse effects they might have.

-   *Abortion*: In many countries, abortion is illegal and carries a penalty of up to several years of imprisonment. However, numerous clandestine abortions takes place every year, resulting in great health risks for the women. Abortions are, of course, undesirable, but making them illegal only forces women to search for risky measures to solve their problem. Changing the law would be desirable.

-   *Away from the health focus towards popular participation*:
    The involvement of the health care system should be limited to the use of contraceptives in the population policy. Often the contrary takes place and the population policy becomes an appendix to the health policy. The results of such programmes have shown to be limited almost all over the world. What is needed is massive support for the population policy, integration of popular participation and strong commitment. Only such an openness can result in the needed decline in fertility levels which again seems to be one of the most important conditions for a sustainable development in the third world in general and in the Sahel in particular.

North East Arid Zone Development Programme NEAZDP

Jens Christensen
NEAZDP, Nigeria

Introduction

The NEAZDP is financed by the Federal Government of Nigeria and the European Economic Community (EEC) under the 6th European Development Fund (EDF) of the Lome III Convention. Borno and Yobe State Governments are the Executing Agency with technical and management assistance from the consultants Danagro Adviser A/S and Ramboll and Hannemann A/S.

The broad concepts and objectives of the Programme are to motivate and to assist the rural population by improving their standard of living through the proper use and management of the natural resources. This can be detailed as follows:

i.   to combat natural- and man-induced desertification processes;

ii.   to increase agricultural productivity through improved and ecologically and technically appropriate farming techniques;

iii.   to change the attitudes to the environment;

iv.   to develop conservation and agricultural extension services;

v.   to expand the socio-economic infrastructure within the Programme
        Area through community based initiatives;

vi.   to increase the technical expertise of the community, States, and
        Local Government organisations;

vii.   to improve education and health of the rural population for
        sustainable development of the communities.

This paper is the result of the work performed by Nigerian and expatriate
professionals and colleagues of the North East Arid Zone Development
Programme (NEAZDP) over a period of four years. The paper expresses
how far we have reached in integrated rural development, aiming at
sustainability, but without pretending that there is no room for far more
academic and practical inputs to improve through more reliable and
accurate data on the still very weak foundation for achieving sustainable
development.

Natural resources and their use and management

NEAZDP completed aerial photography surveys and photo mosaics in 1991
and 1992 which are used for the preparation of base maps on scales of
1:250,000 and 1:50,000.
    Thematic maps are under preparation to cover present land use. These
maps show: grazing reserves, land suitability, livestock movement and
grazing areas, forest resources, drinking and other water resources, settled
human population distribution, local governments, and districts and
lawandoms (group of villages under 1 village head (Lawan)).

The preliminary results with regard to major classes of present land use are
shown in table 1.

Table 1

Land use	ha	%
Upland cultivation (cropped)	374,000	15
areas (fallow)	331,000	14
Rangelands	1,491,000	61
Fadama cropland:		
flooded rice	12,000	
small scale irrigation farming	15,000	1
Fadama rangelands	192,000	8
Oases	6,500	< 1
Settlement sites	3,000	< 1
TOTAL	**2,424,500**	**100**

Human Population

The human population in the about 1200-1500 villages in the rural areas is estimated at some 850,000 in 200,000 households which gives an average of 4.25 persons per settled household. Another 400,000 people live in the four major towns of Nguru, Gashua, Geidam, and Jakusko, which makes a total of 1.25 million.

The net human population increase is currently running at about 2.75% per year. At this rate the number of people will increase by 75% over the coming 20 years to 2.25 million.

Livestock numbers and stocking rates

Livestock husbandry is the most important economic activity both at present and in terms of potential. An average ownership of 2.2 Tropical Livestock Units (TLU) per settled rural household makes a total of 444,000 TLU, with another estimated 70,000 TLU owned by people in the four major towns. This gives a total of 514,000 TLU owned by settled people.

Semi-settled/nomadic livestock owners own an estimated 10-15 times the number of livestock per household of settled people. It has been difficult to establish a correct estimate of the number of semi settled/nomadic livestock owners, but it seems not to be unrealistic to consider more than 1 million TLU owned by the semi settled/nomadic people.

Although the current productivity of the livestock sector is low, an average of 0.2 to 0.3 kilogrammes per day of live weight gain at Nigerian Naira (NGN) 22 per kilo can be considered. This gives an annual rough estimate of the value of meat output in the order of NGN 2-3 billions.

An estimate of the all year livestock carrying capacity is made in table 2.

Table 2

Area	ha(000)	CC(ha/TLU)	CC(TLUs)
Upland fallow	331.0	10	33,100
Rangelands	1491.0	10	149,100
Fadama rangelands	192.0	8	24,000
Upland crop residues	374.0	4.5	83,100
Rice crop residues	12.0	4	3,000
Irrigation crop residues	15.0	nil	nil
TOTAL			**292,300**

Please note that the calculations in table 2 assume uniform carrying capacity in the two rangelands categories, i.e. it takes no account of denuded and bare stretches, particularly in the Manga Grasslands and older alluvial terraces alongside the fadamas. If such areas are assumed presently to cover 25% of the two rangelands categories, their present carrying capacities fall to 111,800 TLUs and 18,000 TLUs respectively, and the total carrying capacity to 249,000 TLUs.

It appears from the comparison of the carrying capacity with the stocking rate that this rate is several fold in excess. The hectares of range land, cropresidues, fodder production, declining rainfall over the years, carrying capacity, and stocking rate are not yet fully clear and need further studies. The present situation obliges us to consider present resource use and land management patterns for livestock as totally unstable in the short term, and therefore present output also unsustainable in the short term.

In Sahel environments, overstocking by a factor of 2 or 3 has been noted in many locations both now and over the last 20 years. In many areas, the livestock sectors, while undoubtedly under stress, have not declined at rates suggested by the degree of overstocking - probably because standard carrying capacity estimates of the type listed above largely ignore the key feed value of browse for livestock. Furthermore, the estimates rarely take into account real changes in species composition of the range. In an overgrazed pasture, perennial grasses are replaced by annual grasses which are in largely resistant to the heavy grazing pressure. In other words, as grazing pressure increases, the decline in carrying capacity is not linear, but descends in stages which may contain niches of lesser or greater sustainability as determined by local factors.

Upland farming

Let's now look at the crop production. The staple food for most of the 1.25 million people consists of millet, sorghum and cowpea. These crops are mainly cultivated in the upland farming areas under an unreliable and unpredictable rainfall without the possibility of irrigation. The ever decreasing soil fertility, erosion, soil degradation and numerous risks associated with the production, result in low yields of about 0.5 tons per ha., with great variations from year to year.

For the above reasons, only an estimated maximum of 374,000 ha out of the approximately 700,000 ha are cultivated each year to provide the staple food and to ensure some degree of food security on which also major parts of the Republic of Niger depend.

With an average daily consumption of grain of 0.4 kilogrammes per capita and assuming that protein, vitamins, and minerals are supplemented by legumes, fish, meat, milk, and vegetables and that export equals import in the area, the annual requirement is in the order of 182,500 tons of grain. A conservative estimate of post-harvest losses of 10% results in a requirement in the order of 200,000 tons of grain which, with the current yield estimates, requires 400,000 ha as against the 374,000 ha actually under cultivation.

Decreasing soil fertility is due to reduced fallow periods, the need for the cultivation of unsuitable land, invasion by cattle on crop lands, rupture of the former good relationships between crop farmers and pastoralists and the unavailability of chemical fertilizers.

The current situation would demand at least 24 ha. of range land to maintain the soil fertility of 1 ha of upland farming area with the current yield levels.

Water resources
The major threat to the improved use of the land resources under the prevailing climatic conditions is, however, the scarce water resources and their inadequate use and management. This problem has been vigorously addressed by NEAZDP over the past 4 years, concluding in the following:

i.   The water of the Yobe-Komadugu river basin should be considered as one system and managed as a whole by a single agency, with all planning and developments considering the benefits to *all* water users within the five States of the basin.

ii.   A minimum of 1,030 million cubic metres of water per annum flow along the Komadugu-Yobe at Gashua, 45% of which is derived from the Hadejia system originating largely in Kano State and 55% from the Jama'are system which originates in Plateau and Bauchi States. The figure for the total volume of water is based on the recommended volume of 1,350 million cubic metres from the National Conference on Water Resources Management held in 1977, but reduced to account for the lower rainfall experienced in the past 20-25 years.

iii.   Releases from the upstream dams, particularly Tiga and Challawa Gorge must be managed and coordinated to fully respect the second recommendation above and to ensure timely flows of water for *all* the various downstream users.

iv.   Further dam development upstream, particularly Kafin Zaki dam in Bauchi State, must be shelved unless a comprehensive economic assessment can prove the economic benefit to *all* downstream water users.

v.   If the Kafin Zaki dam is constructed, then the release mechanisms on the dam must be sufficient to recreate the existing average annual flood on the Jama'are/Katagum/Yobe system.

vi.   Yobe and Borno States will, through NEAZDP, implement the most efficient use of water by all users, from the border with Jigawa State to Lake Chad. All five states concerned must take all necessary measures to ensure optimal use of the limited water resources available.

vii.   To ensure that future developments can be planned with regard to the whole basin, the present hydrological network must be rehabilitated and extended.

Farm forestry

Farm forestry in the NEAZDP area forms an integral part of practically all production systems. But at present, tree cutting for fuel wood, fodder, building construction, fencing poles and several other uses are by far in excess of the annual wood production.

In an area under high pressure of overpopulation (both livestock and human) numerous risks associated with the production are considered as a last priority by the rural people. Risks associated with the production are typically: depletion of soil fertility, threats of increasing desertification and inadequate social-economic conditions, tree planting and care. NEAZDP has, therefore, not so far been very successful in improving on farm forestry.

Fisheries

Fishing in the riverine area of the Programme is significant for cash sales far beyond the boundaries of the programme area and is contributing greatly to the nutritional standards of the rural people throughout the area.

Not much is known about the fishery and the Programme has requested the FAO to look further into this matter in 1994.

Concluding remarks

The four most likely scenarios of the unknown water resources management regime in the future are:

1: No water resources

2: Present situation

3: Bagauda agreement (not respected)

4: Water resources for all land resources (unrealistic)

Even scenario 1, however, must be considered as realistic due to the interdependency of the production systems sustaining the human and livestock population throughout the area. Scenario 4 is likely to be unachievable. The Programme, therefore, will focus its attention on the scenarios 2 and 3 in the coming years in order to ameliorate the production. It is a great challenge to address this state of affairs over the next 10 to 15 years, which is probably the minimum period required to reverse the present trends in their use and management.

Who is meeting that challenge?

NEAZDP aims at achieving sustainability, which is a composite of: *political, institutional, financial, technical, economic, social, and environmental* sustainability.

In a large scale development programme such as NEAZDP the participation and involvement of Governments and Financing Agencies from the onset cannot be ignored. This involvement is necessary to address major policy issues and to provide institutional back up for the grassroots development to be effective in the long term. From the other end, the results of many grassroots developments must be combined to improve on the policies decisions and institutional back up which are needed to support a dynamic integrated rural development.

The ownership of NEAZDP is in principle vested in the rural population of 800,000 inhabitants under the leadership of their traditional ruling system and to some extent with active participation from the 8 Local Governments. It has taken most of the initial four years to create this sense of ownership and hence the responsibility for self-help which goes along with that.

Programme organization
The first level in the Programme Organization is the Federal Government, the EEC, and the Borno and Yobe State Governments. These bodies are represented in the Executive Committee, which gives guidance on policies and major financial issues to Programme Management, which is the second level with autonomy in the day to day operation of the Programme. Programme Management consists of four sub-programmes: Management and Administration, including global services, actual management, financial control, marketing & credit, and monitoring & evaluation; Rural Production, including manpower development & training, primary & adult education, primary health, women's development, natural resource management, production (crop, irrigation farming, livestock, farm forestry and fisheries), and field team operation; Rural Infrastructure, including civil works, rural water supply, workshop & mechanical services, and appropriate technology and; Formal Training in and outside Nigeria.

For the initial period, and still for some years to come, the second level in the Programme Organization - "The Management" has acted as spokesman to promote the right policies decisions and institutional back up and in return use these for grassroots development, which takes largely place through small scale private enterprise and existing village development organizations.

Level 2 in the organization will hopefully be redundant over a 5 to 10 year period and thereafter concentrate its efforts on formal training with a wider geographic coverage beyond the present Programme Area. Levels 1 and 3 will remain with the possibility for political, institutional, and financial sustainability.

Key staff

The focus of attention has gradually become a village participatory development, assisted by Development Area Promoters (DAP), Field Technicians (FT), and Village Development Promoters (VDP).

The DAP is the coordinator - or rural development manager, he is well educated and a native of, and resident in, his area of work. He facilitates the links between the NEAZDP management, Local Governments, other development agencies, traditional rulers, and the rural people represented by their VDPs.

The Field Technicians are the semi-professionals in agriculture, forestry, livestock, fisheries, rural engineering and water supply, veterinary services, primary and adult education, district nurses, women development promoters, primary health workers, and others. They are in part employed by the Programme, but increasingly funded by the Local Governments. All reside in their area of work and assist the DAPs in performing the rural extension tasks in partnership with the rural people.

The VDP is identified, recruited and supported financially by the village community he/she represents and has well defined characteristics and functions.

The above staff presently works in 18 Pilot Villages, 344 Cluster Villages in 18 Development Areas, and three Grazing Reserves.

The approach

The main approach for NEAZDP's rural development is the transfer of funds and power of decision to the rural population for the implementation of village projects. The decisions are taken to the largest possible extent by the villagers. This is done either in informal groups or in the existing groups of the Village Development Associations and their sub-committees.

Until 1990 the Programme Area was lacking financial and technical support. The initial four years were, therefore, needed to improve village water supply, infrastructural improvements, basic skills, primary education, and others, simply to create the necessary self confidence and atmosphere for an active participation of the population in addressing more essential issues for sustainable rural development.

During the same time, the staff has been better trained and has acquired more practical experience in rural development. Finally, the staff has gained more knowledge about how to improve the management of natural resources.

In the future, we will focus on rural extension to improve on the balance between natural resources, population, and production.

On the basis of the initial four years of experience as summarized in this paper, NEAZDP has been entrusted with the responsibility of preparing a long-term strategy around these three basic issues. This strategy, which is nearing completion, will be the subject of a Workshop in February 1994 with participation of NEAZDP staff, as well as Government and Financing Agency representatives. The Workshop will review the major strategy issues, the time frame for implementation and the necessary level of investment and funding.

The level of expectations of NEAZDP staff on the outcome of the Workshop confirms the overall approach for integrated rural development and sustainability towards the year 2000, and from Yobe State Government an institutional strengthening of rural development throughout the State.

Adaptive research

To narrow the gap between the extremes of the present situation and a sustainable development, adaptive research on numerous topics must take place alongside the rural extension and development.

General research already initiated include:

i.   Fisheries biology and technology survey by FAO

ii.   Food policies and nutrition study by the International Food Policy
        Research Institute

iii.   Farm Forestry by the International Centre for Agro Forestry

In addition adaptive research will be within five major areas:

i.   Water Resources

ii.   Soil and Water Management

iii.   Crop, Range, and Livestock Production

iv.   Environment and Natural Resource Monitoring

v.   Socio-Economics

These five areas include about 40 research projects for which outlines have
been prepared by or in cooperation with National and International
Research Institutes, which will implement the projects with its own funding
or under research contracts with NEAZDP.

Le Programme Sahel Burkinabè

Les études de base necéssaires pour la mise en oeuvre d'un programme de développement

Michel Lofo

Planificateur Régional, Burkina Faso

Introduction

Les pays du tiers monde ont expérimenté de nombreuses interventions dont le but exprimé est de leur propulser un développement qualitatif en faveur des populations pauvres. Les interventions sous forme de programme ou de projets ont connu des résultats différenciés.

Dans l'ensemble, la plupart de ces pays, sont toujours à la recherche de leur voie. Le Burkina a traversé plusieurs périodes qui ont été caractérisées par ces programmes. Il y a eu d'abord les interventions directes sous forme de comptoirs Européens, durant la période coloniale, puis des appuis à travers les sociétés privées extérieures, suivis par l'encadrement par les services de l'Etat, pour aboutir aujourd'hui à l'approche ONG. L'approche sectorielle a cédé la place à celle intégrée. Maintenant on préfère l'approche globale avec gestion des terroirs et participation des populations. Les différents changements témoignent du fait que les orientations prises n'ont pas donné entière satisfaction. Les causes de ces échecs sont multiples parmi lesquelles on identifiera l'insuffisante connaissance du milieu dans lequel on évolue. Intervenir sans une étude approfondie du terrain peut entrainer des conséquences négatives sans issues. Mais quelles peuvent être les études de base nécessaires pour la connaissance du milieu avant ou pendant la mise en oeuvre d'un programme de développement? Pour répondre à cette question délicate, nous prendrons l'exemple du programme Sahel Burkinabè. Après un inventaire des études exécutées dans ce programme, nous tenterons de proposer les études principales nécessaires. C'est une démarche osée, difficile surtout dans un pays agricole arriéré où tous les problèmes se posent de manière urgente et prioritaire.

La présente contribution comporte les principaux points ci-après:

- la présentation du Programme Sahel Burkinabè

- les études pour l'exécution du Programme Sahel Burkinabè

- les études de base pour le démarrage d'un projet de développement intégré.

Présentation du Programme Sahel Burkinabè

Le Programme Sahel Burkinabè (PSB) est un programme de lutte contre la désertification et pour le développement. Il s'exécute dans les trois provinces plus au Nord du Burkina: Oudalan, Seno et Soum. Le pays compte trente provinces.

Cette région sahélienne s'étend sur 36 869 km² entre les latitudes 13° et 14° Nord. Les températures varient de 12°C à 45°C selon les périodes de la journée et de l'année.

La population de la région en 1991 était de 611 359 habitants. Cette population qui représente 6,6 % de celle du Burkina a une densité moyenne 14 hab./km², la plus faible du pays. Cependant celle du cheptel est très élevée et la capacité de charge est au seuil de rupture.

Objectifs et volets du Programme

Le Programme Sahel Burkinabè est une initiative du Burkina Faso qui prend en compte les orientations des pays membres du Comité Inter-état de Lutte contre la Sécheresse dans le Sahel (CILSS) sur la lutte contre la désertification. Il a pour objectifs:

- promouvoir le développement du Sahel par la participation et la responsabilisation des populations sahéliennes dans les actions entreprises et dans la gestion de leur environnement;

-      sauvegarder, restaurer et améliorer le capital foncier et écologique du Sahel qui est la base de son développement;

Pour atteindre ces objectifs:

-      le programme met en oeuvre une approche concertée et intégrée qui associe les populations et leurs organisations à tous les stades de chaque action;

-      le programme utilise l'approche Aménagement du Territoire et Aménagement des Terroirs;

Pour ce faire, le PSB a retenu trois volets:

-      Elaboration d'un Schéma Régional d'Aménagement du Territoire, cadre d'harmonisation des interventions,

-      Mise en oeuvre de la démarche Gestion des Terroirs (GT) dans des zones-test afin d'expérimenter l'organisation et la participation des producteurs pour un développement global et intégré.

-      Actions d'accompagnement pour préparer la généralisation des expériences dans toute la région.

Afin d'exécuter les trois volets avec le maximum de cohérence et de succès, le PSB a mis en place une structure de concertation qui va du niveau national jusqu'au niveau village et qui regroupe tous les intervenants.

Les partenaires et l'état d'exécution du Programme Sahel Burkinabè
Parmi les partenaires extérieurs on peut citer, la République Fédérale d'Allemagne, le Danemark, les Pays-Bas, le système des Nations Unies, la Banque Africaine de Développement.

La phase pilote s'est traduite par de nombreuses réalisations socio-économiques dans les zones-test. Elle s'est accompagnée selon les zones et selon les projets par l'élaboration d'esquisses de plans d'aménagement. Au niveau régional, l'élaboration du schéma régional d'aménagement du territoire se poursuit avec un retard dû essentiellement à un problème de financement. Toutes ces activités, tant locales que régionales, ont été précédées puis accompagnées d'études du milieu.

Les Etudes Entreprises dans le Cadre du Programme Sahel Burkinabè

Le PSB a adopté une approche globale qui consiste à faire un diagnostic de la région pour définir un cadre d'intervention sous forme de schéma d'aménagement tant au niveau régional que du terroir. Les schémas à l'échelle du terroir doivent s'harmoniser avec celui de la région, pour constituer ainsi un guide d'actions pour l'ensemble des intervenants.
Pour ce faire, trois types d'études ont été menées:

- les études d'élaboration du Schéma Régional d'Aménagement du Territoire

- les études pour l'aménagement et la gestion des terroirs

- les études sectorielles avant l'exécution de chaque activité particulière.

Les études d'élaboration du Schéma Régional d'Aménagement du Terriroire
Ce sont des études régionales pour les trois provinces. Elles sont au nombre d'onze. Ces études avaient toutes été définies par la mission de formulation du programme. Par la suite il s'est avéré que certaines études existaient déjà mais on n'en avait pas tenu compte, et d'autres se recoupaient. Ceci nous a démontré que l'inventaire documentaire avant toute action est aussi importante que l'action elle-même.

Par ailleurs le principal inconvénient est que les études sur le Schéma étaient coordonnées par un conseil basé à New York qui ne connaissait pas le Sahel. Ceci a causé quelques insuffisances dans la synthèse des travaux du projet. La coordination est donc restée insuffisante. Il est nécessaire pour un programme que les études soient menées avec la coordination de ses responsables qui connaissent à la fois les besoins et le milieu.

Ces études effectuées sont:

- étude sur le régime foncier dans le sahel burkinabè;

- études morpho-pédologiques et d'évaluation des terres de la zone du Sahel Burkinabè;

- l'occupation des sols dans le sahel burkinabè;

- les systèmes de production dans le sahel burkinabè;

- étude socio-démographique des populations du Sahel Burkinabè;

- étude socio-économique des sites aurifères dans les Provinces du Sahel Burkinabè: Soum, Séno, Oudalan;

- inventaire des potentialités hydro-agricoles dans les Provinces de l'Oudalan, Séno, Soum;

- étude sur la foresterie au Sahel Burkinabè

- étude sur les infrastructures, les services et le commerce dans le Sahel Burkinabè

- étude hydrogéologique d'implantation de puits et de forages dans les quinze zones-test du Programme Global Sahel Burkinabè

- enquête territoriale en vue de la réalisation des cartes suivantes: végétation réelle, potentialité hydrogéologique, itinéraire de transhumance dans les provinces sahéliennes du Burkina Faso (Soum, Séno, Oudalan).

Les études pour l'aménagement et la gestion des terroirs
Les principales sont:

- étude socio-économique ou diagnostic;

- inventaire des ressources naturelles comprenant l'occupation des sols et la carte de vérification;

- étude et carte des unités géomorphologiques;

- étude et carte de classification traditionnelle des terres ou carte ethno-pédologique.

Ce sont des études qui ont été menées dans les terroirs par les responsables des projets avec l'appui de consultants externes.

Les études de faisibilité avant l'exécution de chaque activité
Au démarrage du PSB beaucoup d'activités ont été menées uniquement parce que ce sont les propositions des producteurs. Il fallait laisser le paysan décider et non décider à sa place.
    Il s'est avéré, que pour toute activité, il faut une petite étude préalable afin d'éviter, les inadaptations. Cette étude de faisibilité simplifiée définit les modalités d'exécution de l'activité et surtout les conditions de sa gestion, de sa prise en charge, de sa pérennité et de sa reproductibilité dans d'autres zones. Il s'agit notamment des études sur les moulins, banques de céréales, magasin de stockage, boutique villageoise, périmètre maraîcher, embouche ovine et bovine, retenues d'eau etc..

Les études de base nécessaires pour la mise en oeuvre d'un projet de développement intégre

De toutes les études entreprises dans le cadre du programme, cinq semblent indispensables pour le démarrage d'un programme intégré comme le PSB, un programme situé dans une zone essentiellement pastorale:

- l'étude socio-démographique;

- l'étude sur l'évaluation des terres;

- l'étude sur les infrastructures socio-économiques de base;

- l'étude sur les systèmes d'élevage.

Ces études permettent d'éclairer la situation mais elles ne sauraient constituer les seules études. Au fur et à mesure de l'avancement du travail, on identifiera d'autres études complémentaires selon l'orientation des activités.

Une cinquième étude est obligatoire en cours et en fin de projet: c'est l'étude d'impact du projet que malheureusement beaucoup de projets laissent de côté.

Etude socio-demographique

Il est nécessaire avant toute intervention de caractériser la population pour laquelle on veut intervenir, d'identifier ses centres d'intérêts afin de proposer des actions qui répondent à ses aspirations. Les éléments essentiels de cette étude peuvent être:

- *effectif*, notamment l'effectif par ménage pour déterminer la partie active;

- *la répartition géographique* selon les zones écologiques et selon l'occupation de l'espace;

- *la composition ethnique* (régionale, mais surtout villageoise) et l'état des relations entre les différentes ethnies. Dans les zones sahéliennes et en général en Afrique, il existe de nombreuses ethnies dans une même région. Pour le cas du Sahel Burkinabé, dans un même village il y a souvent au moins deux ethnies ou groupes antagoniques. Si ces clivages ne sont pas bien identifiés la mise en place des infrastructures ressucite d'anciennes divisions anéantissant ainsi l'effort entrepris. Ces conflits posent des problèmes de localisation, de gestion et de contrôle de l'activité;

- *organisation sociale*: le pouvoir, la situation des catégories sociales par rapport aux centres de décision: hommes, femmes, personnes âgées ... Une place particulière devrait être accordée à la situation de la femme et de l'enfant. Il est nécessaire de déterminer leur rôle dans la production, dans la distribution et dans la consommation des biens et services, leur degré de responsabilité dans la gestion et dans l'utilisation des revenus par rapport aux hommes. Cela permet de déterminer s'il faut des activités spécifiques pour les femmes ou au contraire s'il faut des mesures de sensibilisation pour une meileure répartition de ces revenus. Parfois, les activités introduites pour l'émancipation des femmes se traduisent par une augmentation de leur charge de travail et une augmentation de revenu pour l'homme seul.

Par ailleurs, le niveau d'organisation communautaire de la zone a une grande importance quant au type d'activités et d'infrastructures à exécuter: souvent on prête un niveau d'organisation très élevée à nos structures locales. C'est à peine si l'on ne dit pas que nos communautés sont socialistes.

Or dans la réalité, pratiquement toutes les unités économiques sont individuelles et non collectives. Seules les ressources naturelles sont communes: terres, mares, pâturages. Cela avec des nuances selon les régions et les ethnies. Ce sont ces nuances qu'il faut mettre en relief pour proposer une gestion appropriée des activités. Cependant, l'entraide, la solidarité sont des qualités réelles et constantes que l'on retrouve dans les communautés locales.

- *le niveau d'instruction*: langue d'échange, langue nationale, langue maternelle. Il détermine l'effort à fournir dans le cadre du transfert des connaissances;

- *le droit foncier actuel*;

- *les activités principales et secondaires*: liées avec l'age, le sexe, l'ethnie et la zone écologique. Les rapports entre les deux types d'activités: Les activités principales doivent faire l'objet d'une analyse détaillée sous forme de filière. En déterminant ses influences sur les autres activités, sur la vie sociale du groupe, on apercevra facilement par où il faudra commencer. Dans notre cas, l'activité principale est l'élevage, bien qu'à l'heure actuelle avec les maintes sécheresses, le sahelien devienne de plus en plus agro-pasteur. Les agro-pasteurs représentent à l'heure actuelle près de la moitié de la population.

Etude sur l'évaluation des terres

Cette étude devrait comporter d'abord une étude d'occupation des sols avant l'évaluation proprement dite. L'occupation des sols détermine la tendance générale de la population pour l'affectation de son patrimoine foncier aux activités primaires qui la préoccupent . Par ailleurs elle détermine l'appréciation empirique des populations vis-à-vis de l'aptitude des sols dont elles disposent.

Les éléments à faire ressortir sont: les zones d'habitations; les zones de cultures et de jachère; les espaces stériles: glacis, effleurement pierreux; les zones de pâturages; les zones forestières; le réseau hydrographique.

Cette étude d'occupation des sols comportera une étude diachronique (étude comparée dans le temps) et permettra par la suite une évaluation des terres (évaluation physique: aptitude des sols et surtout évaluation socio-économique). L'évaluation socio-économique explique pourquoi les producteurs utilisent tel type de sols pour telle spéculation. On devrait alors déterminer les méthodes qui améliorent la production. Pour le cas de notre projet, l'évaluation des terres a éte faite physiquement par le BUNASOLS qui a conclu au fait que les terres ne sont pas aptes aux spéculations actuelles. Il aurait fallu nous dire pourquoi les gens pratiquent ces cultures et enfin comment améliorer les sols pour que la productivité augmente.

Etude sur les infrastructures socio-éonomiques de base
Le niveau des infrastructures (quantité-qualité-répartition géographique) détermine le niveau de développement de la région par rapport au reste du pays. Il détermine par ailleurs l'effort minimal que devrait fournir le projet s'il veut transformer la situation présente de manière positive, sensible et durable. Sans ses infrastructures de base, tout effort sur la production reste ponctuel, partiel et sans impact à long terme. Dans une région comme le Sahel Burkinabè, quelques infrastructures sont particulièrement importantes:

Les voies de communication: Un département comme Sebba est enclavé quatre mois sur douze. Sur les huit autres mois il est désenclavé pour les structures qui ont des moyens particulièrement adaptés (véhicule tout terrain). Dans un tel contexte l'échange avec l'extérieur devient marginal. Les marchés locaux ne sont fréquentés que par les habitants de la zone. Ce qui entraine des problèmes d'approvisionnement.

Les infrastructures d'éducation: école "classique" et centre d'alphabétisation: Le niveau d'information, de compréhension et la capacité à s'instruire seul constituent des éléments clé pour l'acquisition de connaissances nouvelles et pour l'ouverture à des innovations en vue de l'auto-promotion. Ce niveau détermine déjà l'effort à fournir par le projet pour faire passer ses messages.

Les moyens de transfert de l'information et d'animation - les structures d'encadrement et la radio rurale: Si l'on pense parfois aux infrastructures comme les écoles et les centres d'alphabétisation, on pense rarement aux radios rurales locales qui sont des infrastructures légères que l'on peut créer dans chaque zone d'intervention. Elles sont un outil de sensibilisation rapide des masses qui peut compléter le travail de l'encadrement sur le terrain. Les structures d'encadrement doivent faire l'objet d'études pour mesurer leur capacité réelle (effectif et qualité du personnel, sa répartition dans l'espace) et déterminer les principales alternatives à ces structures. Les relations entre ces structures d'encadrement et la population seront analysées afin de déterminer la meilleure voie conduisant à un transfert rapide de connaissances vers les populations.

Etude sur les systèmes d'élevage

Dans chaque région, il y a des activités particulières qui prennent une place importante dans la vie des populations. Elles caractérisent cette région. Pour le cas du Sahel Burkinabè, c'est l'élevage; ailleurs ce sera la culture de coton ou d'igname ou le petit commerce. L'élevage au Sahel influence l'ensemble des actions, réglemente la vie économique et sociale de la société. Il est indispensable qu'une étude sur le secteur soit entreprise afin de déterminer la politique à mettre en place. Cependant du fait des effets de la sécheresse, on assiste à l'heure actuelle au passage d'un système pastoral dominant vers un système agro-pasteur. Cette étude sur l'élevage abordera: les aspects fourragers; les aspects production; les aspects commercialisation; utilisation des revenus de l'élevage; l'impact de l'élevage sur les autres activités sur l'organisation sociale ainsi que le taux de substitution de l'élevage par rapport à l'agriculture ou aux autres activités.

Au Programme Sahel Burkinabè, une telle étude n'est pas faite; on se heurte à un problème de manque de spécialistes.

Jusque-là, les projets se contentent d'exécuter de petites activités d'embouche sans impact réel sur la situation d'ensemble.

L'étude d'impact du projet

Cette étude est aussi importante que l'exécution du projet. C'est elle qui détermine la véritable portée du projet. Elle suppose qu'avant le démarrage, on ait pris le soin d'évaluer la situation de départ, qu'en cours du projet on ait mis en oeuvre un système de suivi des réalisations du projet. On pourra alors mesurer les effets du projet sur son environnement à chaque étape de son évolution.

Démarche pour la réalisation des études

A ce niveau nous voulons évoquer trois aspects:

La conduite des études: L'équipe du projet est indiquée pour diriger les études de base à effectuer. Elle élaborera les termes de référence avec rigueur en tenant compte des complémentarités et des cohérences. Connaissant le milieu, le niveau des informations disponibles, elle guidera avec efficacité les consultants choisis pour les études, elle participera aux synthèses et il lui sera plus aisé de proposer un système de suivi efficient. Elle veillera à ce que les consultants travaillent plutôt en équipe avec elle que de manière isolée et qu'ils travaillent durant la même période. Au PSB il s'est écoulé trois ans entre la première et la dernière étude. On a ainsi rencontré des difficultés de comparaison.

Le niveau des études à effectuer: Pour un projet de développement intégré, ces études de base distingueront trois niveaux: régional, local (zone ou village) et niveau famille. Le niveau régional permet de déterminer des zones homogènes tandis que le niveau local permet de tirer des conclusions pour l'ensemble des zones de même type. Les deux niveaux concourent au choix de la dimension des activités (activités locales, activités régionales). Le niveau famille détermine la qualité et quantité de main d'oeuvre disponible pour les activités. Il détermine par ailleurs la répartition des tâches et des pouvoirs entre les membres de la famille.

Les résultats des études seront synthétisés dans un schéma d'aménagement pour l'ensemble de la région et des plans d'aménagement pour les différents terroirs quand bien même le projet n'a pas l'ambition d'intervenir au niveau régional, cette analyse régionale demeure importante pour situer le cadre général de l'action du projet et déterminer quels pourront être les freins externes du projet.

L'inventaire des travaux antérieurs: Souvent le groupe de formulation n'a pas eu le temps d'inventaire les travaux antérieurs. C'est à l'équipe de projet de s'acquitter de cette tâche avant d'entreprendre les études de base définies plus haut. Il s'agit de faire l'inventaire des études déjà réalisées et qui concerne la zone du projet et de faire un bilan succinct sur les interventions antérieures (projets, ONG) afin d'identifier les acquis et les faiblesses et mieux orienter le travail à venir. Ce travail permet d'éviter l'éternel recommencement.

Conclusion

Le nombre et le type d'études nécessaires est fonction du genre, de l'envergure du projet et de l'enveloppe financière disponible. Cependant, certaines études doivent se faire pour servir de base si l'on veut exécuter des activités d'un impact durable et qui contribuent réellement à un changement positif à long terme. Malheureusement l'horizon des projets (2 à 5 ans), l'exigence de ceux qui financent (résultat concret en très peu de temps) obligent les responsables de projet à négliger ou à simplifier beaucoup d'études.

L'innovation introduite dans un groupe ne peut être acceptée que si:

- elle répond à un besoin réel de ce groupe

- elle s'appuie sur les valeurs de ce groupe.

Seule l'étude patiente et bien ciblée permet un éclairage de la démarche.

The Diffa Project
- administrative aspects

Torben Lindquist
Ministry of Foreign Affairs, Denmark

Introduction

The Diffa Project does not exist and, the way it was intended in the first place, probably never will.

To summarize the history of the project, it began with the staff of Projet Danois, the Danida-funded, rural-water project in Zinder and (to a lesser extent) Diffa. The idea was to establish a local natural resource management project along the lines of the fashionable "terroir" approach. Danida, also keen to try this approach, gave its go-ahead for a set of preparatory studies that were to be designed and organized by I.Krüger, the company running the Projet Danois. The studies were concentrated on a small area in the Arrondissement of Maine-Soroa not far from Goudoumaria.

My two co-speakers of this session (Christiansen, Ørum, this volume) will undoubtedly go into some detail when speaking about these studies. In the autumn of 1992 a Danida mission visiting Niger formulated the project, adding a Zinder leg in and around some villages of the Arrondissement of Mirriah where the Projet Danois had already carried out some minor agricultural activities. The same mission also formulated a separate water project in the Departments of Diffa and Zinder.

In early 1993, an appraisal mission not only turned down the two project proposals, but also left the choice of the future scene of a natural resource management project (terroir style) open, to be established by new studies.

The appraisal mission came up with a project design, the Programme d'Aménagement et de Gestion des Ressources Naturelles (PAGRN) dans les Départements de Diffa et de Zinder, putting together water supply and the development and management of other natural resources under the same umbrella. In addition to water and "gestion des terroirs", the mission included a regional planning component for Zinder, for which a separate request had been received from the Nigerien authorities. It is intended that a future phase of this component make an attempt to bridge the gap between the traditional levels of planning in Niger (i.e., the national and the regional levels) and the dangerously narrow terroir level.

In the beginning of 1994, Danida will recruit an advisor for the planning component (phase 1) of PAGRN, in order to start as soon as possible, while the two other components are being tendered to Danish consulting companies in collaboration with Nigerien "bureaux d'études". All three components will be placed within one Coordination Unit in each of the two departments and a Liaison Unit in Niamey, all manned by Danida advisors and Nigerien counterpart staff. The units will be situated at the respective offices of the Ministry of Finance and Planning, which ministry will also be the general institutional anchorage of the programme.

Two sorts of administrative issues

Two administrative problems will be briefly dealt with: preparing, administering and managing this type of project, and the relationship between the national administration and the project. The PAGRN project will be used as an example.

Project planning and administration

From an administrative point-of-view, the PAGRN type of project is cumbersome and difficult. This stems from the facts that the approach is largely new and little known (at least to Danida); the approach is thoroughly participatory; the general field of activity of the project (environment, natural resource management) is vast, multi-sectorial, without clear-cut and obvious boundaries.

The newness of the approach is reflected in the lack of knowledge and experience in how to handle the pastoral population within the "terroir" approach, in spite of the importance of this part of the population and economy, and in spite of the damage that can potentially be done to it by the "terroir" approach, and the lack of experience and methodology of monitoring this type of project.

With the participatory approach, it is not possible to determine the activities and outputs of the project. Among other things, this includes uncertainties pertaining to the time schedule and the budget. The Logical Framework, that is usually used for project design by Danida as by many other donor agencies, is thus not a really appropriate tool in this context.

The broad field of natural resource management and environmental issues is cumbersome, both because donor agencies themselves have been used to looking at development in terms of individual sectors, and because the administrations of recipient countries, including Niger, are characterized by the same compartmentalization, often to a much higher degree.

In the case of the donor agency, this means a lack of experience in handling projects cutting across traditional sector boundaries. This is so in spite of the presence of so-called integrated rural development projects in most donors' portfolio: such projects were usually clusters of small sector projects or components with varying degrees of coordination among them; they did not attempt what the "terroir" approach tries, viz. to take a global and comprehensive view at the situation of the individual community and on that basis work its way into relevant activities and sectors. And in close connection with this: They were not focused on "demand driven" development to the same extent as the new approach.

Secondly, the donor agencies have few persons (within as well as outside the organization) whom they can rely on to assist in carrying out such projects. In the case of Danida, this problem is aggravated by Danes who often speak French poorly.

The problems arising from the character of the national administrations in the Sahelian and other countries, these will be dealt with below.

In project design and administration, the main difficulty is *flexibility*. This and the uncertainty it implies, are difficult for bureaucracies and donor agencies.

There are four reasons for this difficulty, first, it appears that normal, fairly restricted studies are insufficient to determine, not the need nor the feasibility of the project, but rather its form and direction. Second, the same complexity and vastness of the subject makes long and broad studies necessary as a baseline for project monitoring (see Christiansen, this volume). Third, even after having got the results of the preparatory studies, it is impossible to foresee and program precisely the field activities that will be carried out and the visible results of the project. The time it will take to reach a certain coverage or to achieve certain targets is also largely unknown. Fourth, as the time frame of this type of project is long, this further adds to uncertainty.

The way these uncertainties are being handled in the case of the PAGRN is the following:

- Regarding the *water component*, a 5-year phase was specified, although this component, too, is not without its uncertainties. This is mainly because it is basically going to use the same participatory approach in the needs assessment and investment preparation as the terroir component, and will be able to respond to a much wider range of needs by offering a much wider choice of technology than the present project. Still, it was felt that the things to be done within this component were fairly well known.

- Regarding the *terroir component*, it was essentially assumed that this type of component and approach could be justified and would be feasible in some form or another. Therefore, the over-all project set-up was designed to make possible the implementation of the component over the 5 years of the project for a start. But at the same time, the uncertainties regarding the appropriate approach in terms of geographical coverage, the inclusion of pastoral areas and populations, the choice of technology, etc. were too important to justify a specification of a 5-year project.

As a result, a 1½ years introductory phase was designed, consisting essentially of the preliminary studies deemed required, and of training of staff. By this design it is hoped to involve the local authorities and the local people more intensively than would have been the case if such studies were carried out before the start of the project. Also, this solution will promote accumulation of knowledge about the region and justify the project within the organization itself.

Part of phase 1 would be to formulate phase 2, including the final choice of the initial geographical coverage as well as the strategy of expansion. The proposal for phase 2 would have to be appraised by Danida, and the budget for it approved by the Board of Danida. In order to avoid an interruption of the activities in order to await the approval and arrival of funds for phase 2, a certain amount for phase 2 activities is already being included in the phase 1 budget, to be allocated and used by authorization by the appraisal team.

The transformation of the Nigérien administration

It has often been stated that the national administrations of most developing countries, the Sahelian included, are not geared to tackle efficiently such matters as natural resource management and environmental issues. This stems from the fact that the ministries, each dealing with one particular aspect of what constitutes the much broader field of natural resource management, are often more or less isolated organisms, each one with its own specific sectoral justification, tasks, hierarchical structure and reporting relationships, professional identity and pride, techniques, staff, and set of messages to be extended to the target group. Moreover, these organizations often see each other as something like competitors. Their activities aim at the same target group, whose attention, and obedience, each one of them is dependent upon in order to justify their activities.

The management and staff of such bodies - who themselves come from farming, herding, or similar social backgrounds - and the rural people themselves know well that this sectoral organization of the administration and of the support services does not adequately reflect the reality of rural households and communities and the resources they manage.

The transformation of this system has hardly begun, and apparently more at the initiative of donor agencies than because of internal pressure, as well as under the pressure of economic crisis. The latter seems to be playing two roles: First, it puts the donors in an even stronger position than they already had. Second, it gives rise to the necessity to cut down and to reform public services. This implies among other things: a weakening of the public sector in rural areas; introduction of user payment of part of the support services and of certain investment related costs that had hitherto been borne by the public sector; and increased relevance of private and self-help organizations as suppliers of services.

Different opinions exist of these three changes. Yet they are, no doubt, essential features of the condition necessary to transform the above-mentioned system toward a genuinely demand-driven development of rural support services.

Projects that focus on rural communities' development potential from a point of view of ecological sustainability therefore are in a dilemma: They depend for their success on a set of support services that conceptualize problems and issues the same way as they do themselves, namely from the user side. On the other hand, only their success will bring about the necessary condition for the establishment of such a system, namely "empowered" rural communities.

In view of this situation, any choice of "tutelle" for a project, and of the kind of project/administration relationship in general is risky. This is all the more so in a political context like the one in Niger, where a new constitution has recently been adopted whose consequences for political and administrative decentralization and local democracy are not yet known.

The present situation in Niger, in so far as the transformation of rural administrative and support services is concerned, is far from clear. On the one hand, one of the main political struggles of the new government is precisely with the politically strong civil servants who do not accept the proposed cuts in their budgets wages. For the government, this point is essential, not least in the light of the position of the World Bank for whom such cuts constitute a condition for extending credit to the country. What the precise outcome of the fight will be, cannot be told in advance. Two things *can* be said, however: public employees feel pressured, and in addition they lack the resources required to be productive in their job. Both factors demoralize the civil servants, and only those with a strong personal motivation make an effort to do a worthwhile job.

On the other hand, and specifically regarding the environmental and natural resource management field, two foreign agencies have led a fight against one another that has resulted in restructuring of this field within the administration. What the role of the Nigeriens has been in that struggle is less clear.

The outcome seems to be that the UNSO-sponsored Programme National de Lutte Contre la Désertification (PNLCD) lost out to the World Bank and its Programme Intégré de Gestion des Ressources Naturelles (PIGRN), and that as a result a Cellule de Gestion des Ressources Naturelles (C/GRN) has been established which is attached to the Sous-Comité du Développement Rural (SCDR). Like the PIGRN, the SCDR is placed under the Ministry of Agriculture, although the committee chairmanship is vested in the ministries of agriculture and of water supply/environment in alternation. The C/GRN is composed of a few professionals whose task is to constitute a focal point for natural resource management in Niger and to prepare a national NRM programme.

At the national level an attempt is being made to overcome the sectoral divisions at lower levels. Here, the solution consists of the so-called Groupes de Travail GRN at Département and Arrondissement level. These are composed of local staff of those departments that are most closely involved in natural resource management issues, e.g. agriculture, animal husbandry, environment, water supply.

The concept of such groups is not new: The full team of Directeurs Départementaux already constitutes the COTEDEP, the Comité Technique Départemental, which is a standing committee under the chairmanship of the Préfet. For more specific purposes, such as regional planning, ad hoc groups have been set up. The problem with them is the same: Each director reports to his own sectoral system, and his level of activity within the group entirely depends on the priority given at the national level of his system.

Thus, the problem is both the above-mentioned compartmentalization of the technical services, requiring coordination and cooperation of several systems with different tasks and priorities, and the lack of decentralization making it impossible for local staff to respond to local opportunities and needs. Accordingly, most Groupes de Travail GRN do not work properly, or at all, and their members do not appear to be quite aware of their role.

The Groupes de Travail do not represent an attempt to transform the administration towards being able to handle issues cutting across the traditional sectors in a more integrated and coherent way than before. At most, they constitute the beginning of a long learning process within each of the involved systems and their staff members. More of a potential for genuine transformation can be seen in some of the elements of the Principes Directeurs d'une Politique de Développement du Secteur Rural outlined by the SCDR and adopted politically in 1992.

On the "demand side", i.e., the rural population in its role as client of the administration and the support services, the most important fact is the adoption of the Code Rural. Ideally, this act should eliminate obscurities in the rules regarding the rights and obligations pertaining to natural resources. Among other things, this should reduce the potential for arbitrary rulings by the authorities, and in general increase the scope for local decision-making in these matters. A certain degree of "empowerment" of local communities should result. Another statement in the Principes Directeurs indicates a similar direction, although it is less specific and has been heard for many years in more or less the same terms: The intention to leave more responsibility and influence to rural people than before, when it comes to rural development activities, especially by basing the management and financial responsibility for such activities on the beneficiaries themselves.

On the "supply side", these orientations are matched by an intention - or the sheer need - to reduce the scope of government activity and presence. Most important, private sector actors and self-help organizations are to be promoted. The other element is to allow the technical services of the government to compete with private enterprise for jobs to be done for development projects. This latter possibility is not spelt out in detail, and it is questionable which scope for such competition actually exists. But potentially, it both gives the staff of the public services a chance to do relevant work under adequate conditions, and gives the clients a chance to decide themselves which types of services they want from the public sector, and according to which criteria to evaluate them. Under the right circumstances, this may result in accountability to the clients that never existed before.

All these elements of change may or may not work out the way they should for the rural population. They may even result in further stratification and increased differences of opportunity for different groups.

The project and the Nigérien administration

How does a project of our type position itself vis-à-vis the public sector in order to assist in making this process going the right way?

In the case of the PAGRN, the following choices have been made:

* As the general responsible of the project, the Ministry of Planning was chosen. This ministry's département and arrondissement level offices have a hierarchical edge over the other technical departments because their directors are Secrétaires Généraux Adjoints of the Préfectures/Sous-Préfectures and can therefore act as coordinators of the technical services of the département/arrondissement. In reality, their coordinating and organizing power is not always up to the standards of the formalities, but still, they are in a better position to perform such functions than the other ministerial units.

    Another advantage of this ministry is that it does not have any extension staff in the field. Thus, it is not biased in favour of the use of and support to any particular technical staff.

* Especially within the water component, but perhaps also of some relevance to the "terroir" component, the fact of choosing le Plan as the general institutional basis of the project should also facilitate a shift from the present situation where technical departments not only implement or organize investment decisions, but also take such

decisions themselves, to a situation where such decisions are taken by a non-sectoral forum, i.e., ideally by people themselves under the general umbrella of le Plan. It is important to understand the local priorities. These should define the fields of operation, but as an action of local planning under the general umbrella of le Plan.

* Unlike the water component of the project that - for its investment implementation activities - will be based at the two Directions Départementales de l'Hydraulique, the "terroir" component of the project will operate largely independently of the technical services, in that it will recruit its own staff to carry out the basic work. This makes it largely autonomous from the administration. It gives it the opportunity to use the services of the various technical departments in a selective way, and when required to enter into the above mentioned type of contractual relationship with them. (For the reasons developed above, it may be an advantage to promote this type of relationship even in cases where it would have been possible to make the services work without contract and payment.)

    What this amounts to is empowering the clients of the project vis-à-vis the administration.

* To the extent that the local technical departments are to play specific roles within the national natural resource programme (for the moment headed by the Cellule GRN, later probably a revised structure supported centrally by a coming World Bank project, that Danida intends to co-finance), the project will, of course, acknowledge such roles and support the departments in them. Within the present set-up, this will mean supporting the Groupes de Travail GRN at both département and arrondissement level in their monitoring and guiding roles. In the former role, the groups will be supported by financial means to carry out monitoring and supervision, and regarding the latter the two groups at département level will be prominent parts of the advisory committees of the project.

*　　Furthermore, to await a clearer situation regarding the development of the public and the private sectors and the way their utility for the rural population may be promoted, the terroir component of the project has not yet been formulated in any detail. Only a preparatory phase consisting in the set of studies has been specified.

Basic Information Requirements for Project Preparation

Sofus Christiansen

Institute of Geography, University of Copenhagen, Denmark

Roughly speaking, the process of creating any project will pass through three phases. In medical terms they might be termed: diagnosis of the disease, prescription of cure, and - finally - the treatment. Failure at any stage may lead to disaster. Likewise, the proper preparation of the project is decisive and hence sufficient knowledge as a prerequisite.

Most scientists would agree on the desirability of having all necessary knowledge at hand before any project is suggested. Realities are, however, different. The high cost and duration of scientific investigation necessitates to renounce on ideals and try to make the best from what is available.

Existing Information

Available information on most developing countries is of two types:

-      standard information of a general kind

-      specific information, either sectorial or regional

Standard information usually comprises:

a) *Information generally collected*, independent of any Danida inspiration or request:

- Statistical data, more or less according to UN-specifications. Additional data ordered from national bodies.

- Various planning documents, often World Bank-inspired.

- Topographical maps at a scale seldom better than 1:250.000 except for selected areas, generally often at even smaller scales (world coverage exists only at 1:1.000.000). In general, the maps need up-dating and correction. Often also geological maps exist.

- Partial air-photographical coverages, rarely recent except where engineering/project purposes has covered expenses. Scales vary greatly and are usually not better than 1:50.000. Also quality is variable and usually low, since much photography has been done with used war-time equipment.

- Satellite-imagery coverages of varying resolution, quality, and age.

b) *Information collated at Danida initiative*:

- Situation-and-perspective analyses.

- Country profiles, country analyses.

- Programme and project identification reports.

- Feasibility reports on proceeding projects.

- Project documents of proceeding projects.

- Reports on studies related to specific planning and project problems.

- Collected materials on projects of other agencies (eg., from UNDP, WB, NORAD, NGOs.)

Although the lists can in some cases be extended, they may equally well be even more meagre. Anyway, even if the lists given above are an indication of what normally exists, the information is usually only available with great difficulty and often only in the capital of the developing country. The standard statistical data may be unavailable because it is out of print or kept by local authorities unwilling to let anyone have access to them lest they be lost. The problems of getting adequate map and airphoto-coverages for field work is well known. In spite of too low resolution, the lower (but not exactly low) price and recentness of satellite imageries make these a preferable alternative to airphotography, especially so if the multi-band properties of satellite radiometric scanning can be utilised.

User's problems with the available information can usually be listed under three captions:

- tracing

- acquisition

- usability

Tracing is an enormous problem in most developing countries because of lacking registration and collection of materials. Very often what actually exists on a given topic is locally unknown. Often information can only be traced with assistance from locals with a good memory or via foreign citations. Not even the highest authorities in charge can usually assist unless the material wanted is pre-identified. Although less pronounced, a similar problem may exist within Danida.

The problem of tracing and identifying existing information can be overcome by the establishment of registers and libraries. If these are supplied with electronic search-systems based on both topic, region and key-words, they can be extremely effective. Such registers have been established in a few developing countries but are still lacking in most as well as in Danida. Apparently Danida's archives are rather meagre in spite of the tons of reports, documents etc. every year brought back from missions. Possibly the reason for the poverty of information is that so much work is now carried out by external consultants. A rule for depositing all materials collected during contracted Danida-work might be helpful; consultants might still be given permission to copy for personal use if they so wanted.

Even the best registers need to be supplemented by an 'institutional memory'. This is often amazingly effective in developing countries. Even when personnel has been promoted or removed, it is often possible to identify knowledgable persons for finding or identifying missing materials. Within Danida the 'institutional memory' seems to be in a process of deterioration. One of the reasons is a system of job-rotation, meaning that most of the overburdened employees will stay only for a period of one or a few years with the same job before being transferred to remote destinations - detrimental for building up of a store of personal knowledge and for the establishment of relations of mutual understanding with the clients in the developing countries!

As may be gathered from the above, the tracing problems are much the same whether standard or Danida-collated information is being sought after - although they vary to some degree.

Acquisition of traced/identified materials is a different story altogether. Generally Danida conducts a commendable, open policy of giving access to its materials and making it available to those needing them. In developing countries there may be hindrances because of policy or merely from practical reasons. Sometimes information is kept back for security reasons, but in a few cases possibly also because it is felt that 'knowledge is power'. Good personal relations often overcome such problems, but genuine practical problems may persevere. Most national and some international published sources of information are printed in a limited number of copies, and photocopies usually are difficult to produce in developing countries. Again, some type of center for distribution with attached copying facility could solve the problem. At the same time, such a center could be instructed as to what should be accessible and what not, which may also pose a problem.

Even when the difficulties in identifying and acquiring needed data have been overcome, all problems are not solved: those of sufficiency of information (coverage and quality) remain. Quality control is difficult and time consuming, and in some cases even impossible. Checking with other accessible information (eg. by 'triangulation') is often the only way out, but not always sufficient. It may be necessary to collect or create new information.

Information collected for specific purposes:

At the beginning of more important projects, lack of specific information may be so evident that something is done to overcome it. Usually only information specific for further operations would be considered, while other ones may await project implementation before being collected. The reason for waiting is, of course, that information needs tend to become more and more specific and well defined during project operation, and hence more easy to cover efficiently.

Present Danida policy allows studies to be attached to projects under-way, which has meant much improved chances for projects to be operated on a solid knowledge-base. The analogy to shallow-water navigation comes to mind: in such a situation frequent sounding and control of bearings are recommended. Even by cautious sailing, disaster may strike, showing the initial charting has been insufficient.

The big problem is to judge when the initial knowledge is sufficient for the safe formulation of programmes or projects, and the necessary information that remains to be acquired can safely be so during programme or project operation. Such a risk assertion is in no way easy, especially since conditions for any programme/project are often of a dynamic rather than a static nature.

The Zinder-Diffa example

As an illustration of the problems encountered in providing sufficient knowledge for the preparation of a programme the recent Zinder-Diffa programme shall be referred to as a background in the following.

In late 1992, Danida decided upon an intensified intervention in the Republic of Niger, specifically in the two easternmost departments, Zinder and Diffa. One project, the Projet Danois worked by Krüger Consult for almost 15 years, should enter into a new phase; another proposed project for strengthening administration in planning especially in its environmental aspects was about to be formulated, but had in the drafting stage met stiff opposition from the Nigerien side.

The problems pertaining to a third project of the programme were not less important. This project aims at improving rural livelihood by organising the use of village territories - in accordance with fashionable views based on the 'terroirs villageois' concept. The Danida staff in Niger had collected a considerable amount of relevant literature, visits were paid to Nigerien projects dealing with it or similar ideas, and - rather unusual, but encouraging - a series of scientific studies had been conducted on important aspects relating to such a project. The deciding to make this investigation deserves appreciation, and the critical comments mentioned below are not meant to scare Danida from even more extensive use of preparatory studies. Rather, pointing out weaknesses is done to pave the road for future improvements.

The reporting of the preparatory studies was encompassing with an analyses of special topics such as:

- natural resources, sand encroachment, millet fields, localisation of cuvettes (i.e. small, basin-like depressions in the landscape), monitoring of changes (esp. on mobile sands and millet fields), landscape profiles.

- pastoral resources

- pastoralism

-     woody resources

-     water resources

With the reporting came also a volume of synthesis. Finally, a case study throws light on the resource system of a village and the social context in which it functions.

It remains laudable that basic studies of such a quality and comprehensiveness were made available at the very beginning of a programme formulation, and with the additional information collected, the formulation of a project document would appear almost problem-free. That was not quite the case, and the major reasons why are given below.

The most important was perhaps that the analyses did not give convincing indication for the selection of the cuvette area as the most suitable one for a project. An underlying assumption for the selection had apparently been that the area had under-utilised water resources and thus was one of notable promise, distinguishing it from most others in the two eastern departments. The cuvettes were, however, found to have no proven water resources of any sizeable potential for development, at least no more than many other parts of the departments. Although some of the cuvettes proved to have ground water of a useful quality, the total storage was not adequately assessed and the rechargeability of the resource not proven. (Additional research was carried out, but results were difficult to use because the series of hydrological observations, eg. on recharge, are too short to allow safe conclusions). As a consequence, the risk for failure of a cuvette-based project was considered too high, and the later search for project location was therefore not restricted to that of the cuvettes.

The more generally applicable findings of the studies had also some shortcomings of a general type which shall be mentioned here:

-     the land use records are difficult to interpret because too little is generally known on farmers' land use strategies in relation to factors such as seasonal distribution and quantity of rainfall. This sometimes makes interpretation of land use mapping a guesswork: is it expanding or is there just a change of location because of a special pattern of rainfall?

-      utilisation of pastures, of course, depends on available biomass. But
       it also depends on palatability, digestibility and accessibility of the
       vegetable material. For a planning of pastoral uses the analysis at
       hand did not seem sufficient.

-      the system of land rights was not, and is not, well enough known
       for an understanding of constraints in the present use of land.
       Possibly a full perception will require insights in the historical
       development of land uses, which is not too well known presently.
       Actual land occupation and land use could easily be of a rather
       temporary kind reflecting an imbalance in the wake of the serious
       droughts with very specific problems. Many former nomadic
       herders have thus become settled agriculturists, but will they remain
       so if they gather enough livestock?

Some of the questions raised may need answers before the launching of a
project (eg. the one on farmers' strategies seems basic for an interpretation
of mapped land uses), others may have answers already. A tougher one is
the need for clarification of the 'terroir villageois' concept. Is it true that
the sedentary villagers are the best administrators of 'village lands'? In an
area of utmost importance to the nomadic herders at least two parts should
be involved in the administration. As a matter-of-fact, definition of 'village
land' remains a problem.

Conclusion

Generally seen the two questions to be posed in relation to the building of
the knowledge-base seem to be:

-      what information needs to be acquired? and

-      when is the information needed?

The information-need clearly depends on both the topic of the project and its relatedness to other topics (i.e., to the project's purpose and scope). The detailing of both will usually increase throughout the total project cycle; hence the information-need is growing over time. But since the information available for satisfaction should always be greater or equal to the need, it is an integral part of all projects to make sure that information is gathered well ahead of needs. Unless the aims of the project are well defined early enough to allow this operation to take place, serious delays or mistakes could happen.

A very high responsibility lies with the early project identification, the problem diagnosis. If this is well done - which requires information of a rather broad type emphasising linkages and proportions - accumulation of information can better keep up with the growing needs, because these tend to become more specific and 'sectorial'. In the case cited above, a major problem was that the initial information was collected too narrowly and become irrelevant with the final project formulation. The data collected were too predetermined and became unaligned to the needs of both the diagnosis and the later project formulation.

The information also became incoherent. Apparently, the various researchers mostly worked independently, which made it difficult for the later coordinator to amalgamate the different part-analyses into a whole.

The conclusion from this experience is therefore not that it was erroneous to conduct a series of pre-project studies, but rather that the studies must be closely connected to the various stages of development of the future project to be cost-efficient. An optimal solution might possibly be to let project formulation develop along with and in constant dialogue with studies in the preparatory phase - so far the different time-schedules allow this.

Similar ideas, are practised by Danida in the form of 'the rolling project development'. By this the project initially is governed by relatively broadly formulated intentions; these are gradually being specified and translated into plans for activities and attached work plans and budgets - along with the accumulation of the necessary basic information.

The Diffa Study - An example of Gestion de Terroir Approach?

Thorkil Ørum
Krüger Consult, Denmark

"Gestion de Terroir" Approaches
- Main Issues and Problems

Gestion de Terroir is the right approach

There is a widespread conviction that Gestion de Terroir (GT) approach is the right approach. This conviction, however, is based on faith rather than evidence. In fact, so far we do not have sufficient evidence from GT projects to assume it works.

Considering the history of development paradigms, where magic solutions have been put forward numerous times, it might be advisable to be more modest. But the GT approach is attractive because of its insistence on and integration of resource management, participation, decentralization and empowerment/transfer of rights, e.g. hot key words of the 90ies.

Main issues

GT is an integrated approach to development combining the environment system, the socio-economic system and the legal system through effective participation.

Underlying assumptions - horizon 2030-2050

It is relevant to remember that the insistence of the GT approach for development is based on a number of implicit assumptions:

1.   The Sahel will remain a rural region mainly depending on agricultural production. Production can be maintained or raised through better management of natural resources and simple technical improvements.

2.   Climatic conditions will remain broadly unchanged. A more arid climate will decrease agricultural production. A more humid climate would diminish the need for careful management of natural resources (soils and water).

3.   Democratization processes including decentralization and participation demand stable political conditions, clearly defined and enforced legal procedures and a benign state.

The Diffa Study

Background
Krüger Consult has been responsible for activities in Zinder and Diffa for more that 15 years. These activities were financed by Danida. First, the activities concerned water supply, both urban and rural, later also in rural development. The rural development component evolved from off-season gardening (fr.: culture de contre-saison) to micro-projects (fr.: micro-réalisations) including dune stabilization, irrigation, tree-planting, and organization and education of beneficiaries. All projects cooperated with local extension services. It was found that scattered activities in rural development were not effective; therefore, an integrated approach was proposed.

Goudoumaria, which covers 4,300 km2, was chosen as a pilot zone. The choice was based on previous work in the zone.

Danida requirements and objectives
Krüger Consult had great liberty in defining study contents. In fact, Danida only insisted on defining an appropriate institutional linkage with the Nigerian administration.

Study Contents
The feasibility study for the GT project was composed of 8 studies:

-      Mapping of natural resources

-      Hydrology

-      Agrostology

-      Organizational study of exploiters of natural resources

-      Study on natural resource exploitation

-      Pastoral systems

-      Analysis of development projects

-      Study on fuel wood

Eight reports and one synthesis volume were produced.

Evaluation
The organization and coordination of the execution of the thematic study components were not optimal.

There were no -or very little- participation in study by future beneficiaries.

Studies were maybe too general and not very operational, partly a consequence of the fact that no commitments for implementation of activities were made.

Principles for GT projects

Broader framework: Do not plan a GT project without being sure that some general conditions are fulfilled or will be successfully addressed during project implementation, such as access and traditional rights to natural resources.

Participation: No projects without effective participation! Provide effective participation in all phases. A number of techniques have been tested and proved their usefulness: RRA Rapid Rural Appraisal), PRRA (Participatory Rural Appraisal) etc. This will certainly require a different project design (new cycle).

Remember that there may exist conflicts between different groups, which have implications for a participatory approach.

Partners: Who decides? Development for whom? These questions are not new, but still highly valid.

Classic GT project design:

PROJECT PREPARATION

▽

PROJECT DOCUMENT

▽

STUDIES

▽

ACTIVITIES

▽

MONITORING/EVALUATION

In the classic GT project design there is little or no implication of future beneficiaries in project identification and preparation of project document. Many resources are often used on project preparation and studies. The time span from project preparation (first contact with the beneficiaries) and implementation of activities may be considerable.

An alternative GT Project Design
An alternative GT Project Design could be as follows, supposing that baseline studies are available:

FORMULATE PROJECT DOCUMENT:
FLEXIBLE FRAMEWORK
∇

METHODOLOGIES FOR
PARTICIPATORY ACTIVITIES
∇

DEFINE AND IMPLEMENT A LIMITED
NUMBER OF CONCRETE ACTIVITIES
CORRESPONDING TO REAL NEEDS
∇

PARTICIPATORY MONITORING
AND EVALUATION
∇

FURTHER ACTIVITIES

FURTHER STUDIES
(As appropriate)

RESOURCE MANAGEMENT PLAN

Briefly summarized this alternative will imply some major constraints as well as benefits, such as:

Problems:

-       Conditions for commitment of funds

-       Management and Staffing

Advantages:

-       Trust may arise between project and beneficiaries

-       Learning by doing.

Integrated Development Project in the Seno Province, Burkina Faso

Inge Schou [1]
Water & Power Planners, Denmark

Presentation of the Séno project

What follows is a brief description of the PSB/UNSO project, (here called the Séno project).[2] This project, financed by Danida and executed by UNSO, is one of the projects under the "Programme Sahel Burkinabè" (Lofo 1994, this volume). The Séno-project started its first 5-year phase in September 1991. It covers Séno Province in the northerneastern Burkina Faso. It is a natural resource management project based on the approach "gestion des terroirs".

By the end of 1991, the project team started working in Sebba department, where 3 tests zones were chosen (Kabo-Gountouré, Lac Higa, and Sud-Est (Tantiabongou, Pansi, Bira, etc.)). Each zone includes several villages or quarters. Later on in 1992, another 3 zones were added in the northern part of the province (Falagountou, Tibabé, and Koria).

The main activities concern environment, livestock and agriculture. The project team mainly national experts, extension workers etc. The team works in close collaboration with local extension services and the other projects included in the "Programme Sahel Burkinabè".

---

[1] Associate expert at the PSB/UNSO project from May 91 to May 93. In this presentation the studies are looked at from the point of view of a team member.

[2] PSB/UNSO = Programme Sahel Burkinabè / United Nations Sudano-Sahelian Office.

Project Studies and Research

The studies carried out within the framework of the Séno project during the two year period, where I worked at the project, are the following:

* History of the Yagha area (today the department of Sebba).

* Socio-economic analysis, Sebba department.

* Land tenure systems in the department of Sebba.

* Development of a methodology for natural environmental monitoring in the province of Séno.

* Diagnostics and recommendations on the utilisation and commercialisation of livestock and meat in the provinces of Séno and Oudalan.

* Socio-economic analysis of the northern part of the Séno province.

* Experiences with the forage plant "bourgou" in the lake Higa, Sebba.

* Suggestions for a soil and water conservation strategy in the department of Sebba.

* Agrarian dynamics and the approach "gestion des terroirs": agropastoralism in the department of Sebba.

The comments to these "classic studies" can briefly be summarized in the following main points:

* In general, the studies have been useful - however to a different extent.

* The majority of the studies have been carried out by external consultants hired on short term.

* Almost every study has been carried out by one consultant.

* In most cases the consultant has only been in direct contact with the project managers. Other members of the team have not been very much involved in the studies apart from a de-briefing before the departure of the consultants.

* The human resources, which exist internal in the project team, have not been used in a satisfactory way. The daily work has priority which means that the team members seldom find time to carry out studies (except from the socio-economic analysis).

* The people who live in the area concerned have not been much involved in the studies.

In addition to these studies the project team has carried out some research activities concerning:

* Cultivation of forage plants in Higa lake.

* Improving milk and meat production by feeding cows "tourteau de coton" (a kind of oil cake made from cotton).

* Growing Acacia albida in Sebba.

The socio-economic analysis

Among the studies mentioned above, I have chosen as an example the socio-economics studies for further discussion. The objects of these studies were firstly, to collect information and analyze the socio-economic situation in the area; secondly, to use these basic information as a frame of reference for the subsequent monitoring and evaluation. The study in the Sebba department was more extensive than the study in the northern part of the province because the existing information available on Sebba was limited.

   The socio-economic analysis was based on a large questionnaire written by the project manager (rural economist), and the extension workers (enquêteurs). The extension workers filled in the questionnaires in collaboration with the villagers. In the case of Sebba, the project manager did the socio-economic analysis on her own. In the second study concerning the northern part of Séno, several persons employed by the Séno project (two extension workers (animaters), one Ph.D. student, the project manager, the sociologist, and the anthropologist) took part in the analysis. The rest of the team was not included in these studies.

Three different questionnaires were used:

*   a group questionnaire (concerning information about the history of the village, ethnic groups, infrastructure, relations to other villages etc.),

*   an individual questionnaire for women, and

*   an individual questionnaire for men.

In Sebba 32 villages were included in the study (319 men + 88 women). In the North 11 villages (138 men + 80 women).

   The socio-economic studies have provided a general view as well as basic information on the selected area. The quality of the information depended on time and human and economic resources at disposal, but the results of the studies were not optimal for many reasons such as:

* The questionnaire and most of the analysis were elaborated by only one staff member.

* The analysis based on the questionnaires (mainly quantitative information) was not followed up by a more qualitative analysis (for example semi-structured interviews with "key persons").

* The analysis showed differences in the production systems between the ethnic groups (rimabé, peuhl, gourmanché, songhai, gaobé). This information was not always taken in consideration when planning activities.

* The extension workers who carried out the study in the field were not well trained to do the work, which among other things resulted in misinterpretation of certain questions and inaccuracy in some answers.

Studies based on a participatory approach

Besides the traditional way in carrying out studies (fr.: études/diagnostics externes), the Séno project has carried out studies based on a more participatory approach (fr.: études/diagnostics internes). These studies have been carried out after the external studies and have included representatives from the villages, the whole project team, as well as the extension services covering the Séno province. In each zone several 3-4 days meetings have taken place.

During these meetings the group of peasants, team members and extension workers have discussed themes such as:

* delimitation of the zone (walked around in the area to identify and map the limits between the different zones);

*   quantification of the natural resources which exist today and calculation on the needs today as well as in 20 years;

*   identification of the users of the zone (sedentaries + people passing through the zone (transhumance)) and their inter-relations;

*   the management of the natural resources according to the usual practices;

*   identification of the existing conflicts in the area and how the local people resolve these conflicts;

*   identification of the main development problems in the area and their relative importance;

*   identification of potential future activities carried out by the villagers in collaboration with the extension services and the Séno project, and calculation of the costs of these activities.

All information was recorded on big pieces of paper, which remain in the zone so that the villagers can use them whenever they want. On the basis of all the information collected during these meetings and discussions in every village/quarter, a management plan has been elaborated for each zone, where the Séno project intervenes.

Studies carried out in this way involve many people, who are concerned about the interventions - from villagers to extension workers to the project staff.

In the concrete case the internal studies/diagnoses were carried out in a period of only 3 months, which didn't give the villagers the necessary time to think about and to discuss the elaboration of the management plan. The internal diagnosis was speeded up, because the project manager wanted to bring the diagnosis to an end before the rainy season and before the external evaluation was carried out by consultants hired by the local government, UNSO, and Danida (evaluation mi-cours in October 1993). During some of the meetings in the villages the discussions were too prolonged and detailed, having as result that the villagers couldn't keep concentrated on the discussions and fell asleep or complained about the way the meetings were carried out.

In conclusion: participatory studies can give very important information and encourage the participation of the population, the project team, and extension workers; but it depends on to what extent the participants master the participatory methods. One should consider these kind of studies as an important supplement to the more "classic" way of doing studies, and not as a replacement of these studies.

Comments on how to improve the studies related to development projects

The experience gained from the activities related to the 8 Séno projects can be summarized in the following main points:

* Involve local people as far as possible; base internal studies/diagnoses on participatory methods.

* Coordinate studies carried out by the local extension services, different projects, local and international research institutes. Start every study by making a survey on existing studies concerning the same subject or region and find out what kind of information is needed and who might be interested in using the results from the study. (A local documentation center could play an important role in this coordination of interventions).

* Integrate the studies as an activity in the project planning. Reduce the number of studies carried out before the project team is installed, and improve the integration of the different people concerned directly by the interventions in the studies (for example recruit the team members as early as possible in the project phase and carry out different kinds of basis studies in the first year of the project before starting a lot of other activities).

* Make better use of human resources, which exist on the project level (let the team members carry out more studies themselves, if they have the necessary qualifications and/or increase the integration of the team members in the studies carried out by external consultants).

* Have in mind that the studies have to be translated into concrete actions. From a project point of view, a study is good if the results can be used in practice.

* Reinforce the research which is based on concrete actions (action research).

Summary of Discussions on the Seno Project

Kjeld Rasmussen
Institute of Geography, University of Copenhagen, Denmark

Inge Schou started out by giving an overview of the many studies carried out during her two years of work in the DANIDA-financed UNSO project in the Seno Province of northern Burkina Faso with emphasis on the socio-economic studies. Several studies were carried out by short-term consultants and involved little interaction with project activities and staff, and thus had limited impact.

The first theme of the discussion was the concept of 'gestion terroir' (GT), in general as well as in relation to 'integrated projects' in Burkina Faso in particular. It was stated that in the present context GT projects emphasize 'people' not just 'land'.

The second theme was the need for carrying out studies, or data-collection, before and during projects. Most contributors emphasized the need for integrating data-collection in project activities in order to ensure that they are closely related to ongoing activities of the project, attain the necessary depth and point towards actions. This also calls for data collection taking much longer time than the 'hit-and-run' studies often carried on by outside consultants. A key factor is the extent to which the experience of project staff is utilized. Some contributors even questioned the very idea of carrying out data collection before the start of GT projects. Lars Engberg suggested that starting a project with one year of studies, carried out mostly by project staff, before any actions were taken, would be an interesting experiment. It was pointed out that the Seno project is nearly an example of this. However, Inge Schou stated that a project needs to demonstrate its relevance locally by carrying out activities rather than just asking questions. A balance must be found.

'Participation' was then intensely discussed. Again, several speakers stressed the need to strike a balance between listening to people and still maintaining the priorities on certain development objectives, as determined by the overall aims and resources of the project in question. Per Christensen stated that participation of local people in the decision-making process is a way of introducing democracy from 'below', rather than from 'above' as it is often done at present.

Finally, monitoring and evaluation was discussed in the context of the flexibility of the GT projects, such as the Seno project. Ideas of integrating monitoring and evaluation in project activities and of involving laymen, rather than just experts and administrators, were put forward. Holger Koch-Nielsen stressed the need to monitor changes in people's minds and perceptions, rather than just external, physical parameters.

Michel Lofo summed up the discussion by stating that the main axes of GT-projects may often be defined by preparatory missions having only few weeks at their disposal. This implies that there is a great need for continuous data collection as part of project activities, and that these needs should be discussed with donors, which is often possible in the case of bilateral projects. Donors and projects must state their objectives clearly, rather than mislead local authorities and people about freedom of choice. What is meant by 'monitoring', what to monitor and how is often unclear and must be discussed and decided upon in early phases of GT projects.

Getting Beyond Preparation

Christian Lund
International Development Studies
University of Roskilde, Denmark

Estragon: *I'm tired. Let's go.*
Vladimir: *We can't*
Estragon: *Why not?*
Vladimir: *We're waiting for Godot*
Estragon: *Ah! What'll we do, what'll we do?*
Vladimir: *There's nothing we can do.*
Estragon: *But I can't go on like this*
Vladimir: *Would you like a radish?*
Estragon: *Is that all there is?*
Vladimir: *There are radishes and turnips*
Estragon: *Are there no carrots?*
Vladimir: *No. Anyway you overdo it with your carrots.*

(Beckett)

Apology

As a novice in the arts and crafts of project preparation and design I might in the following apply terms which are not the professionally correct. I hope the reader grant me tolerance. Furthermore, some of the finer subtleties in project preparation process probably have escaped me, and some of the parties who have been deeply involved in the process may find my interpretation of it incomplete and unjust. I admit that a more forthcoming interpretation could probably also be produced. The problem of getting beyond the state of preparation, however, remains.

Introduction

Preparing an integrated rural development project can be an awesome task. There is hardly an end to the studies and questions to which it would be comforting to know the answer in advance. Furthermore, there is a tradition for blaming mistakes in project execution on insufficient preparation. Another rationale is often at play: Since integrated rural development projects invites numerous and detailed studies, and since productivity and productivity increase is often quite modest in the Sahel, a project has to be large in order to justify the costs of merely preparing it. This again often leads to massive, ambitious preparation which at times seems grotesque vis à vis the concrete activities of constructing tube wells, improving the cultivation of date palms, improving animal husbandry. Considering that farming systems change over time and are supposed to, due to project impact, and that specific project objectives often undergo change as project personnel learns more about the priorities of the beneficiaries, very detailed preparation seems labour of Sisyphus.

All this is not to say that an integrated project should not be prepared in advance. It is merely to say that the objective should not be to prepare as much as possible but as little as necessary in advance. Of course, the donor/project needs some notions about who the target groups are and how the society is organized, what the land and natural resources are like and how the production systems are organized. This is well described by Michel Lofo (this volume) and I shall not contest it in detail. Only, there is an inherent risk of "overkill" due to an ambition of pin-pointing The Problem before getting started. Extensive preparation in this type of project may contradict the entire project philosophy of a flexible and participatory approach. A more modest ambition of gradually encircling the problem area alongside with setting concrete project activities in motion might prove more fruitful. Studies should primarily constitute an integrative part of an iterative planning process. Not be an isolated hit-and-run-operation. This, however, requires fairly stable general objectives on the part of the donor.

Furthermore, if large projects require massive justification, maybe one should consider making it small to begin with to leave more scope for manoeuvre. A crucial element in this type of project is thus the transformation from a micro scale experimental activity to activities at a scale which is not insignificant while maintaining close dialogue between project and beneficiaries. Furthermore, the back-up of the project in terms of response to requests for specific studies seems essential.

Outline of the Diffa-project preparation

The process of preparing an integrated rural development project in Diffa is in fact rather paradoxical but illustrates my points.

Briefly, the Environment-activities started out as a relief effort in the wake of the serious drought in 1984. A national strategy of resettling drought-refugees and establishing off-season gardening projects was formulated by the Nigerien authorities. Danida let an existing water project run by Krüger Consult administer some funds for this purpose in the area of the project. During the first years, the objectives were modified. Initially the production of vegetables was advocated for the purpose of improving the diet of the displaced people: it proved much more reasonable for the people involved in the project to produce for the market. The target groups also changed somewhat: instead of addressing only refugees, people who lived in the villages participated. Some experiments of coordinating environment and water components in setting up tree nurseries was tried but largely failed. Over the years, the strategy was changed from setting up off-season gardens to provide technical assistance to already existing production sites. This furthermore entailed identification of other related problems in the project areas. This included e.g., improvement of date palm cultivation, efforts to stop sand encroachment of the production sites. And as the project became better known to the local authorities and the population, other activities were begun such as establishing fire breaks and fixing sand dunes.

It might be argued that this project was a mess. That objectives were constantly changed and that results were meagre when compared with the scale of the problems and the costs of the operation. I would argue, however, that the change of objectives and strategies was a healthy sign of adaptability and some dialogue with the target groups. The irony is, however, that the ability to manoeuvre relatively freely probably was based on the fact that the "Environment component" was considered an accidental excrescence of the Water Project. When reviewed and evaluated, the project was seen as a water project and the Environment component as a kind of experimental laboratory, where considerable freedom was granted to learn from mistakes.

The blissful and secluded existence of the environment component was disturbed when its activities was identified as a future Integrated Natural Resources Management Project. While the activities should continue for some time as the Environment component of the Water project, they also inspired an autonomous larger scale natural resources management project. There was a general consensus among the project personnel, Danida, and external consultants that the knowledge to get started was too limited and a strategy was needed. Since 1990, a number of strategy papers - preceded by strategy missions - has seen the light of day. An example is IIED's three volume Study to Enhance Natural Resource Management in Zinder and Diffa from 1990. This was followed by at least two strategy proposals by Krüger Consult, one of which I produced in 1992. This again was followed by first one mission and report by external consultants in 1992, and then again by another team in 1993. Parallel to these activities two missions made efforts to strengthen the local administrative capacity to develop environment programmes.

How different have these reports then been since it has been necessary to produce at least five strategy papers? Well, they differ somewhat in level of abstraction; some are mostly policy papers while others enumerate the cars and motorbikes necessary to carry out the operation. It is thought provoking, however, that all reports argue for a learning-as-we-go-along-approach. The art of muddling though. This entails that no specific objectives can be established by a mission but must be developed in collaboration with the local people. Thus no specific activities can be pre-determined - we have to ask people what they want. It is thus not a matter of contents that provoked the series of reports.

What has then been the effect of the strategy papers? Well, it would be wrong to say none at all, but one can hardly argue that they have yet resulted in project action on the ground. However, it resulted in a kind of selection of a zone d´intervention around Goudoumaria. Oddly enough, while the strategy papers stimulated the choice of a project zone, it is not easy to find clear strategic reasons for the specific choice of Goudoumaria. The reason which seems to emerge from the strategy papers is that the Environment component of the Water project already had success in the area. Each strategy paper thus gradually took it more and more for granted, while no strategic criteria seem to have been applied in a selection procedure. Consequently, the choice of zone could still be questioned and take the process back to square one.

Another result of the strategy papers was the execution of a number of sector studies such as cartographic mapping of landscape types, and study of tenure systems, studies of the aquifers in the oasis. All studies are most likely very qualified. But how well do they answer the questions that the project will ask? That is very hard to say, but a few problems can at least be identified.

One problem has been that the studies were parallel; not mutually influential or informative. They did not enrich each other. This is because the studies did not arise from concrete problems in the execution of the project. There is therefore a risk that they will only serve marginally when the project eventually will be executed. They may not even address the problems that arise 6 months into the project, since the objectives, the specific sites and the involved populations may change during the course of the project. Data not utilized is data better not collected. If this was the end of the ritual, things could then start. The parties involved in the preparation of the project are, however, painfully aware that some knowledge still is (always is) lacking and complementary pre-studies remain lurking in the wings.

The crucial question is, of course, how much firmer is the ground for the project with the strategy papers and sector studies?

Let me briefly return to a common denominator of the strategy papers. All advocate an iterative, learning-by-doing, ad hoc, flexible, adaptive, participatory process with constant interchange of information between the project and the beneficiaries. This should entail a demand driven formulation and re-formulation of objectives and thus studies as the needs emerge throughout the running of the project. In other words, recommendations concerned the way studies should come about, and not their contents. It is tempting to argue that the long preparation process contradicts the very philosophy of the strategy advocated during it. And it is ironic that while the project was still merely the Environment-component of the Water project, some of these qualities of participatory problem identification, constant interchange of information etc. were present. Not perfectly. Not constantly. Not unambiguously. But still to some extent.

Why, when all strategy papers advocate getting started and then integrate the studies as the project unfolds, have each strategy paper caused its own replication? Clearly it is not caused by the potential beneficiaries. They do not participate much in the initial rounds. It is not their demands that drives the process. Otherwise, we have three groups of actors: Danida, Krüger Consult who has managed the Environment component and wishes to continue to manage the project in its new form, and the external consultants of which some are also anxious to manage the new project when it starts. The latter two thus compete for the favours of Danida. This is not always an easy task, because what does Danida want? A series of Danida policy catch-words such as participatory, demand driven and flexible seems to collide with the unavoidable demand for pre-established justification. In order to allocate millions of kroner, Danida needs good justification. That is only reasonable. This requires an idea of what the problems are and the vicious circle starts again.

Professionally, it is also not always easy to decipher Danida´s policy. Constant change of Niger-desk officer and Danida-in-house-consultants who all seem anxious to start anew by formulating a new project, makes it difficult to get beyond the threshold of preparation. The large number of people who are half involved in the project preparation keeps the policy priorities in a constant quivering flux. With one, pastoralists have a special protagonist, with another the democratic aspect of the project, with another women´s role in the project has special priority and with another institutional sustainability is the key word. Both Krüger and other consultants try to cater for the taste they believe to be particular in vogue at any given moment. While both Krüger and the other consultants have an interest in eventually getting the project off the ground and compete for it, they are not entirely uninterested in the process of preparing it. Thus the pressure for getting started is not strong. The only ones pressed for time are the future beneficiaries ...

Post scriptum - post festum?

Parallel to these efforts, Danida is reconsidering its strategy. Concentration on programme countries and sectors within them seems to be on the agenda. One consequence might be that regional strategies like the present ones for the Sahel, Central America and the SADEC countries will not find room in Danida's policy. Two scenarios: one with and one without a Sahel strategy. If the second scenario prevails, a country like Niger could witness a phasing out of project activities and not the launching of new commitments from Danida. History, however, makes it tempting to operate with a third more ambiguous and capricious scenario. Absence of regional strategies makes it legitimate for Danida to terminate activities in non-programme countries where activities have been limited. For non-programme countries with a substantial portfolio and relative success, however, we could face a phasing into discreet status quo. This would, however, entail an extended policy-limbo for Niger, where audacious experimental projects like the integrated natural resources management project in Diffa will have low priority.

References

Chambers, Robert (1983) Rural development - putting the last first. Longman, London, 246 p.

Maxwell, Simon (1986). "Farming systems research - hitting a moving target". World Development vol. 14, no. 1. pp. 65-77.

Summary of Discussions on the Diffa and NEAZDP Projects

Birgitte Markussen
Department of Ethnography and Social Anthropology
Aarhus University, Denmark

> "...*when I was a serious young student in London I thought I would try to get a few tips from experienced fieldworkers before setting out for Central Africa. I sought instruction from Haddon, a man foremost in field-research. He told me that it was really all quite simple; one should always behave as a gentleman. My teacher, Seligman told me to take ten grains of quinine every night and to keep off women. Finally, I asked Malinowski and was told not to be a bloody fool.*"
> (Evans-Pritchard 1973:1).

Preparation, and the problem of getting beyond preparation itself were the main issues raised by Christian Lund (this volume). In the introduction to this workshop he stated that preparation for the Diffa project has been very long - and probably too long. He also brought up the question of *who* feasibility studies are made for. Are these studies written to satisfy the donor or for the benefit of the future project? Hans Jørgen Lundberg (Danida), explained that the most important reason for the long process of preparing the Diffa project has been changes in Danida policy regarding integrated rural development projects.

It was the consensus of the workshop participants, however, that a preparation phase should be short. In the case of Diffa, it was also mentioned as a complication that the work had been done by a private company. Thus, in general, it was suggested that time can be saved by

making project documents flexible and general rather than detailed and
narrow. It was mentioned that instead of asking how much information is
needed before a project can start, one should ask how little is needed.

If projects are prepared on a general level concerning the approach and
the overall objectives, it might be viable to make integrated rural development projects more process-oriented: oriented towards the local needs and
the participation of the population involved. Jens Christensen mentioned
that much of the success of the NEAZDP project can be ascribed to the
brief outline they had when they started. NEAZDP has instead spent 3
years to map the local resources and to identify what can actually be done
in the programme area.

Monitoring and evaluation was also discussed. It was argued that it
seems as if more and more people are concerned about less measurable
results such as change of behaviour and attitude among the local population. Participatory evaluation was mentioned as an excellent way of
monitoring a project at the village level. Video and radio programmes were
also discussed as alternative ways of initiating internal and external
evaluation processes.

Review missions were also addressed. There were different opinions on
the relevance of short-term review missions. Some believed they were too
short, whereas others thought the missions should be considered as an
opportunity for a project to have an uninvolved observer. A third opinion
was to understand internal and external monitoring and evaluation as
complementary activities. Concerning Danida evaluations, most participants
agreed that Danida should do their own evaluations. Danida should not
depend upon private consultants to analyze projects.

At the end of the session, the future of the integrated rural development
projects was dealt with. Torben Lindqvist from Danida stated that
integrated rural development projects are not the strategy of Danida right
now. He argued that these kinds of projects are very often difficult to get
beyond preparation because of what he called: "*the constrains of the local
structure*". So-called 'mono'-projects, dealing with water, trees etc., are
the priority of Danida for the moment.

Unfortunately, the workshop did not have time to discuss the mono-
strategy of Danida. Thus, in order to avoid being *a bloody fool* as the

British anthropologist Malinowski, put it, development, in the broadest sense, should always implicate an integrated cultural and political process. This is the lesson learned from development projects, especially Diffa and NEAZDP.

Reference

Evans-Pritchard, E.E., Hilary 1973. Some Reminiscences and Reflections on Fieldwork. in: Journal of the Anthropological Society of Oxford, 4 (1) pp. 1-12.

Integration of Research and Development Activities: The Norwegian Sahel-programme

Sissel Ekås
Ministry of Foreign Affairs, Norway

Introduction

Thank you for inviting me here to share with you the experience we have gained in trying to integrate research with development activities within the context of the Norwegian Sahel-programme. Let me right away, however, warn those of you who came here expecting me to reveal some Norwegian magic formula for bringing about such integration that you will be disappointed. We are still struggling to find the right mechanisms for this. It is therefore with a certain humility that I have accepted the invitation to speak to you.

The Norwegian Sahel-programme has been characterized rather bluntly as a costly adult-education programme for Norwegians, but also as a unique attempt to ensure that an often isolated academic research milieu contribute directly to development. A report by an independent Danish evaluation team issued a year ago concluded that the programme was a great idea, but a lost opportunity. Although individual projects had made valuable contributions in their respective fields, the full potential of what was intended to be a coherent programme had not been realized.

Not surprisingly, the prospects for realizing some of the intentions inherent in the original programme concept are better today than at the start of the programme. I shall come back to this later, but let me start with the origins, objectives, and basic approach of the Sahel programme in general, and the research component in particular.

Origins, objectives, and approach

Frankly speaking, the Norwegian Sahel programme came about, not as a result of sound scientific and technical advice, but primarily as a political response to a public demand for action, mobilized by intense media attention on the severe drought and famine that struck the Sahel in the early and mid-1980s. The idea was that the programme should serve as a complement to the substantial Norwegian emergency assistance to the Sahel over several years. So, what was intended as a longterm development programme was initiated in what was still very much a crisis situation demanding quick action on the ground. This was clearly not an ideal context for research.

The overall objectives of the programme were:

*   to improve food security in rural areas and local food production;

*   to improve the management of the ecological resource base.

The basic programme concept was to achieve the overall objectives by means of a coherent and interdisciplinary programme, consisting of well-co-ordinated and complimentary forms of assistance. The underlying assumption was that a programme approach would result in a certain synergy effect, as compared to simply funding a number of scattered individual projects.

Since Norway had no official bilateral representation in any of the Sahel countries at that time, the idea was to make use of more indirect funding channels, such as Norwegian private organizations (NGOs) and international organizations, coupled with support to research collaboration between Norwegian and Sahelian institutions. Thus, these three forms of assistance or channels are often referred to as the three programme components. While any Sahelian country in principle could benefit from assistance through international organizations, it was considered necessary, essentially for capacity reasons, to concentrate assistance through Norwegian organizations and institutions to three countries: Ethiopia, Mali, and the Sudan. These countries were chosen largely because some Norwegian organizations and/or research institutions already were established or had experience from working there.

By the end of 1993, eight years after the establishment of the programme, total expenditures had reached NOK 1.2 billion. More than 100 small and larger projects under the auspices of approximately 40 different organizations had received support. The research component accounted for approximately 8% of total expenditures. Research funding has been split more or less equally among Norwegian and Sahelian institutions.

The research component - objectives and principles

It was acknowledged that the limited expertise available in the mid-1980s in Norway about the Sahel in general, and about the interrelationship between food production, population, and the environment in the Sahel in particular, constituted a major obstacle to the realization of the programme. Also, the available capacity was not well institutionalized. Originally, the idea was therefore to use part of the funding available to support the development of existing research capacity and infrastructure in Norway, which would then assist in the planning and implementation of development activities.

After lengthy discussions of research strategy in various committees, a general agreement evolved that the research component should have a dual objective of research capacity-building both in Norway and in the three focal countries. A third objective would be to generate results that could be used as tools for planning and implementation of development activities, especially those under the auspices of Norwegian NGOs. It was recognized from the very beginning that the major problem would be to find mechanisms by which research results could be translated into action on the ground. In addition to the objectives, the guidelines for research programmes also specified that research should be

-     applied (problem- and process-oriented),

-     interactive (research topics should ideally be defined jointly by researchers, development agencies, and the target population);

-     descriptive (rather than developing new data bases, existing information relevant to resource utilization and setting land use priorities in the target areas, such as soil and water surveys, inventories of indigenous plant species, livestock assessments, wildlife counts, demographic research, mapping, etc. should be collected and presented in a user-friendly way);

-     interdisciplinary (joint teams of natural and social scientists),

-     institution-based (collaboration between Norwegian institutions and partner institutions in the three focal countries).

The focus on applied research of direct relevance to Norwegian development agencies involved in the programme, implied that administratively defined conditions with regard to the geographical focus of the research and its objectives and principles were linked to research funding. The nature of the services originally demanded from the research component was closer to that of short-term studies and consultancy work, rather than longterm research generating new insight, that may or may not be of immediate practical value. Considering that the research programmes with one or two exceptions had to be developed from scratch, what was expected from the research component was a tall order.

Lessons learned

Naturally, it took some time to develop joint research programmes responding to the before-stated objectives. Most of the programmes became operational only in 1988-89, or even as late as 1990. Thus, when an evaluation of the first five years of programme operations (1986-1990) was conducted, it was premature to evaluate the results of the research component per se or its impact on the programme as a whole. Nevertheless, the report concluded that the contribution of the Sahel programme to research capacity-building in the Sahel was by no means marginal. The support to teaching and training of a number of researchers in data collection and analysis, as well as the provision of modern equipment, was very valuable to the partner institutions involved in the programme.

The evaluation report also contains several observations regarding the objectives of the research component and the planning of the programmes. Clearly, the Sahel programme met many constraints. Some were inherent in the programme concept and should probably have been foreseen, while others were more unpredictable in nature. Let me mention a few:

* Time (e.g. disbursement pressure, researchers' other obligations, quick results);

* Administrative/political aspects (e.g. shifting institutional framework, shifting priorities/"zig-zag" policy, staff changes, different procedures, etc.)

* Communication (e.g. problems due to language difficulties, cultural differences, lack of telecommunication means, etc.)

* Logistics (e.g. lack of cars and other equipment for fieldwork, lack of information processing equipment, research libraries)

* Security (e.g. threats to personal security of researchers in the field, theft of equipment).

I should mention here that the evaluation team did not visit the Sudan. The longtime collaboration between the University of Bergen and the University of Khartoum, which existed long before the Sahel programme, has been evaluated separately by another team. The conclusions of this evaluation are quite favourable, which is largely explained by the fact that the Sudan research programme evolved gradually as an equal partnership, motivated primarily by mutual academic interest.

Before I go into the lessons learned, I should like to recall that it is easy to be critical with hindsight. All parties in the Sahel programme surely did their best under difficult circumstances, and on the basis of the knowledge and the possibilities they had at the time. Although it may be premature still to draw definitive conclusions, some general observations can be made at this stage:

*   given the circumstances, the programme concept was clearly too ambitious.

*   the difficulties of co-ordinating a cluster of separate projects spread over a wide geographical area, and carried out by a range of institutions and agencies with widely differing traditions, interest and procedures, were underestimated.

*   the objectives of the research component were unclear and mutually conflicting.

*   funding alone is not a sufficient basis for development of research collaboration.

*   disbursement pressure, coupled with problems of identifying suitable partners, made the planning process rather donor-driven, and also left little room for the intended dialogue between researchers and development workers.

*   true partnerships cannot be imposed by a third party, but should grow naturally out of a joint perception of mutual interests and benefits - i.e., researchers must have something to offer before other researchers or development agents will find it worthwhile to cooperate.

The future

In the budget for 1994, Norway's commitment to continued support for development activities aimed at improved food security and natural resource management in the Sahel is reconfirmed. The total allocation to the programme for 1994 was increased by NOK 10 million to a total of NOK 148 million as compared to the 1993-level. Revised guidelines are under elaboration and will probably be approved early February.

As for the research programmes, we have seen some positive developments recently. More and more research results are now being published and a number of M.A. candidates have completed their degrees; a few Ph.D. candidates are well into their studies. With results becoming available, contacts among researchers and development agencies have also multiplied. The potential for cross-fertilization is now much better than at the outset of the programme, because researchers now have some results to show. The evaluation process also stimulated dialogue among the various partners in the process and as such, served a very useful purpose.

At present we foresee a two or three year extension of on-going research programmes in Ethiopia and Mali. Unfortunately, the situation in the Sudan forces us to phase out the programme there. Some elements of the previous research programme will now be integrated in a regional programme, also involving Eritrea. We also need to clarify the criteria for financing the involvement of Norwegian institutions and alternative ways of securing funding of administrative costs and salaries, to ensure continuity on the Norwegian side. The whole research administration in Norway is in the middle of a reorganization process, that may also affect the institutional framework of the programmes in the Sahel. Of course, the future of the entire Sahel programme will depend on the general situation of the Norwegian development budget.

Just like the population in the Sahel must learn to live with drought as an unfortunate, but given factor, it seems that the research institutions depending on public funding also must learn to live with administrative constraints and shifting political priorities!

Independent Research Components in Swedish Development Assistance

Reidar Persson
SIDA, Sweden

Introduction

In "Sustainable management of renewable natural resources - Action Plan for SIDA 1992" the following is given as an important task in country programmes:

> "Monitoring Environment being a complicated matter, a more thorough follow-up of the effects of projects is necessary. In countries where SIDA is financing major natural resources programmes, a component for monitoring could be financed from special funds. Through experiments, long-term follow up and action research, such components could help to improve the quality of natural resources programmes. It is important that also these components are based in national institutions."

This idea comes from experiences with the Bai Bang project in Vietnam. The Bai Bang is a pulp and paper mill constructed with Swedish assistance. In the early 1980s the possibility of a wood-shortage was brought forward. Work therefore started in assisting Vietnam to establish pulp-wood plantations. Many questions were raised: Was agro-forestry a possibility? Could farmers and co-operatives be given incentives to plant trees? How was the social situation of the local population?

The project resisted these research ideas. They argued that the project should only deal with questions of direct interest to the pulp and paper mill. Questions of marginal interest were not the responsibility of the project.

The forestry part of the Bai Bang project has now been changed towards "social forestry". Wood for Bai Bang is of secondary importance. Today it would have been of great value if we had answers to some of the questions raised in the mid 1980s.

Perhaps large projects should concentrate only on their main task. The need for research can, however, be dealt with by independent research projects. Certain aspects of this issue will be discussed here, yet it must be emphasized that here is only discussed independent research components of value to rural development.

Research in Swedish development assistance

In Sweden traditional development assistance is handled by SIDA whereas SAREC handles assistance to research. SAREC often supports research councils in developing countries. The kind of research being supported is in principle decided by the countries. SAREC rarely points out to the countries which research field is of "Swedish interest". SIDA's programmes in different countries have hitherto normally been decided after a discussion with the Ministry of Finance or the Ministry of Planning.

SAREC and SIDA often work in the same countries. Normally, SAREC and SIDA cooperate with different organisations[3]. So it may happen that SIDA supports agriculture in a country while SAREC supports medicine. There has rarely been correspondence between the programmes of SAREC and SIDA. It has sometimes been a gap between the responsibilities of SIDA and SAREC.

SIDA has sometimes had a limited interest in research that has been considered the task of SAREC. When approached about research ideas

---

[3] What is given here is the historical back-ground. A number of these things are now changing.

SIDA has normally referred this matter to SAREC. This has changed during the last ten years. SIDA now supports a lot of research.

Special programme for research and developments of methods (FOM-programmes)

SIDA supports a number of FOM-programmes. Examples are the following:

- Forest, trees and people

- Farming systems research

- Programme for agricultural technique

- Aquaculture

- Forest regeneration in dry areas

The basic aim of these programmes is to develop and spread new knowledge in developing countries. Some of the FOM-programmes have been very successful. We have, however, a problem in how to use the knowledge gained and introduce it in rural development projects. It is normally an advantage if FOM-programmes are carried out in co-operation with international agencies. In this way valuable results can be spread better.

"The forest and environment programme"

In 1986 SAREC was given 100 million SEK during 5 years by the Swedish Government in order to support research in the environmental field. This was a little unusual as it meant that SAREC should take the initiative to start a certain type of research in developing countries.

As a part of that special programme, attempts were made to complement certain SIDA projects with research components. Projects were initiated in:

- Ethiopia

- Nicaragua

- Kenya

- Tanzania

Short comments to these attempts will be given here. A few comments will also be given concerning experiences in India and Vietnam.

Ethiopia
In the mid 1980s SIDA started to plan support to a rural development project in the hunger-striken Wollo-province. No research component was included in the project. SIDA approached SAREC concerning support to a complementary research component.

A work-shop was arranged in Ethiopia in 1988 to discuss possible research in Wollo. Four research projects were planned but only two were started. These two projects concerned:

- successions of vegetation in "enclosures"

- people and soil conservation (Agricultural Development-based Conservation)

Due to the war, the Wollo-project was closed down. The research components could therefore not support the ongoing practical activities in the way planned. However, a new project in Wollo is being discussed at present.

Nicaragua

SIDA has supported a forestry programme in Nicaragua since the early 1980s. Due to the war, the programme has been fraught with difficulties.

When SAREC's Forest & Environment Programme started, Nicaragua was chosen as one of the countries in which to work. Attempts were made to complement the forestry programme. The components described below were planned jointly by SIDA and SAREC.

To IRENA[4] support was given to research agro-forestry. ISCA[5]/UCA[6] was given support to research dry forests, rain forests and agro-forestry.

It was basically the research on rain forests that complemented the SIDA-supported programme. A part of the SIDA-supported programme was located in the rain forest zone but little was known about the ecology of these forests. The research programme with ISCA and UCA tried to give more in depth knowledge in this respect. The agro-forestry part, was among other things, trying to find methods that could help to improve shifting cultivation and extensive grazing in the "fronta agricola".

CATIE[7] in Costa Rica was given support to assist IRENA, ISCA and UCA in the research. The idea was that CATIE should have research projects in Nicaragua in which the Nicaraguans could participate. The results in the beginning would basically come from the work of CATIE.

A number of Nicaraguans were sent for training to CATIE, but recently CATIE has experienced a number of difficulties, which have influenced the project.

When this project started, Nicaragua had no basis for research. The project can basically be seen as a project for building research capacity, and it will take a long time before research results of great value can be produced.

---

[4] Instituto Nacional Nicaraguense de Recursos Naturales y el Ambiente.

[5] Instituto Superior de Ciencias Agropecuarias. Now it is called UNA (Universidad Nacional Agraria).

[6] Universidad de Centro America.

[7] Centro Agrónomico Tropical de Investigación y Ensenanza.

Kenya

SIDA has supported soil conservation in Kenya since 1974. In the mid 1980s the need for research became evident. There was also a need for more personnel with a greater knowledge of soil conservation.

SIDA initiated support for research on soil conservation at the Kenya Agricultural Research Institute (KARI). That organisation works with implemented research.

SAREC started in 1989 to support a regional M.Sc. course in soil and water conservation at Nairobi University. Around 30 students have been trained so far. In the beginning it was mainly a question of training, but now more and more research of value is coming out.

Figure 1 gives the components of the Swedish-supported soil and water conservation in Kenya. New components have been introduced when need has arisen.

Tanzania

The Kondoa area is one of the most badly eroded areas in Tanzania. SIDA has long supported soil conservation in the area (HADO-project). In 1979, 100 000 cattle were removed from the area. The problem was moved somewhere else! Since 1979 the environment has recovered. There are now discussions about bringing the cattle back.

When the cattle were removed in 1979, no research component was attached. After several years this was deeply regretted. There are a number of things that can be learned from what has happened in the Kondoa-area.

A seminar, financed by SAREC, was held in 1988. A research programme has developed which is now being financed by both SAREC and SIDA. The start up was rather slow but now the programme seems to be running well. For the time being, the following researchers and institutions are engaged in the programme:

-       one Swedish and one Tanzanian coordinator

-       8 Tanzanian, 8 Swedish and one British researcher

-       four Swedish and two Tanzanian research institutions.

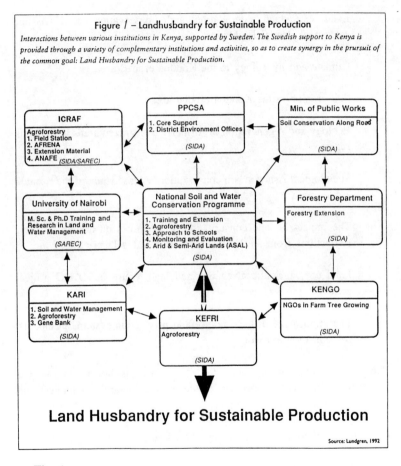

Figure *1* – Landhusbandry for Sustainable Production

Interactions between various institutions in Kenya, supported by Sweden. The Swedish support to Kenya is provided through a variety of complementary institutions and activities, so as to create synergy in the prursuit of the common goal: Land Husbandry for Sustainable Production.

Land Husbandry for Sustainable Production

Source: Lundgren, 1992

Fig. 1

Reenberg & Markussen

The following components are now part of the programme:

- Climate change and landscape development in central Tanzania

- Land use, land degradation and conservation in Mwisanga catchment, Irangi Highlands

- Climate and hydrology in the Kondoa eroded area

- Soil erosion in the Kondoa eroded area

- Ecology and regeneration of formerly overgrazed areas in Dodoma region

- Sociological issues in soil conservation: A case study of the Kondoa eroded area

- The impact of land conservation programmes on human fertility: Reflections from the HADO areas in Kondoa District

- Land use, agricultural change and land tenure in HADO project area

- The importance of local tenure and production systems in modern soil conservation work

The research component being described above is not in any way a component of the HADO-project. However, the research component and the project do influence each other.

Furthermore, there are some closely associated projects:

- Rainfall partitioning and dry season water flow from springs in semi-arid Tanzania

- Studies on the ecology and regeneration of formerly overgrazed areas in Dodoma with respect to soil conservation. Parallel to this there is a research component which is more directly involved in the HADO-project and which deals with e.g. stall-feeding.

India - Tamil Nadu

SIDA supports a couple of large social forestry programmes in India. One of these is situated in Tamil Nadu. When the project was being planned, universities in India were used for pre-studies. At a later stage they were used for monitoring.

After several years, the need was felt for more independent research. Large funds were invested in the projects, and it was necessary to get a better understanding of the effects.

In 1989 a research NGO was established (Society for Social Forestry Research and Development, Tamil Nadu). One research officer is attached to the coordinating unit of the project. The NGO handles a research fund which is used to support research about social forestry. Researchers can apply for funds and the decisions are taken by a board with Dr.Swaminathan as the chairman. The board can also take initiatives for research.

Thirty-five research projects have been funded within the fields of germplasms, growth promoters, fodder research, agro-forestry models, and studies of policies and planning. The funds from SIDA amount to about 1.7 mill SEK per year.

Vietnam

In Vietnam SIDA still has a forestry programme which aims at the development of forestry in five provinces in the north. There are, of course, many uncertainties about what are the best things to do. In Vietnam there are a number of competent researchers when given funds, they can come up with valuable research but, cooperation with Swedish researchers would, however, be of value.

A research fund is now being discussed. Most likely a council will be supported which will take decisions about support to research.

Conclusions

Many good reasons for independent research components can be given. In cooperation programmes personnel on both the donor and recipient side change very often. An independent research group will often be engaged for a long period of time and can have an institutional memory and ensure that results/reports of interest are not just forgotten and that results of scientific interest are published in scientific journals.

Research can easily become "academic" but if it is carried out in close contact with practical work this risk diminishes. Research can tackle questions of great practical importance.

Now there are normally three actors in large projects - the country, SIDA and the consultancy firm. An additional independent knowledgable voice, without self-interest, would be of value. If we had had an independent research component in many of SIDA's rural projects, necessary changes and adjustment could have been made earlier.

It is, however, necessary that a research group shows discipline. The decisions, right or wrong, must be taken by bureaucrats. A problem is that professors (and generals and ambassadors) believe they know everything about everything. They rarely do!

What we have learned is that development of an research component requires a lot of work. The start-up phase is time consuming. Results will not come fast. It can, of course, go faster if expatriates are used but the final result of such a donor-driven approach is doubtful. In a longer context it is better if nationals are heavily involved from the very beginning. "Do not make research for us without us".

It is probably an advantage if there is some direct Swedish involvement in the research, as in Tanzania. The problem is that this "expatriate research part" can become very costly. As a rule the costs for the expatriate research component should not be more than 50% and the expatriate personnel must not dominate. The ideal form is probably to establish a fund for certain types of research. A council takes decisions about which research projects to support. Some kind of Swedish/expatriate involvement is, as mentioned, of value both when it comes to research and the research council because we have a self-interest to get some Swedes involved and expatriates can also see things with fresh eyes. In certain countries a research council may be difficult to establish. In such cases it may be necessary to take a decision about supporting a "research package".

The really difficult question in all this is how to identify the important questions. We have often tried workshops. It can work but it normally takes a long time to develop a programme. It is also possible to work with research councils, advisory groups or ideas of individual researchers.

Every situation has to be discussed on its own merit. Probably it is best to have a mixture of workshops, advisory groups etc. This is really a problem. We should, however, keep in mind that we often have the same problems when it comes to identifying important research questions in developed countries. Sometimes our attempts in Sweden run into difficulties and there is no absolute truth.

Development assistance is in reality often trials on a rather large scale. In industry such type of activity would be very research-intensive. This is not the case, however, in development assistance. As rule of thumb programmes to support rural development and natural resources management ought to include a research component. This research component should be 5-10% of the project cost.

Research Components in Bilateral Development Projects - Danida policies

Klaus Winkel
Ministry of Foreign Affairs, Denmark

Danida has not as yet developed specific policies in the field in question, but the issue has been discussed by Danida on several occasions. What follows is the main lines of thinking in Danida with regard to research components in bilateral development projects.

In principle, Danida encourages integrating research in projects that have other objectives. Danida feels that research promotes the main objectives, and that they can mobilize and strengthen local research - often in interaction with Danish researchers.

We have rather few examples of project documents and budgets where research is singled out as a specific activity being implemented with the mainstream of the project. One significant example is the large primary health programme in Tamil Nadu and Madhya Pradesh in India. Here, money has been allocated for the involvement of an Indian research institute with the purpose of conducting operational research. The research was to be used to monitor the progress and impact of the programme. The fate of this component illustrates the problems confronting research in development cooperation. In spite of the fact that the funds had been allocated and could not be used for other purposes, the Indian authorities never initiated this activity - in spite of Danida encouragement. The administrators in the Indian health administration apparently did not see the need for research - to them the problems were straight forward. Danida also suspects that adding research activities to what is already a complex undertaking was not appealing to the local administrators.

Another more successful example, is a primary health care project in Guinea Bissau, where for several years Danish medical scientists have conducted longitudinal epidemiological research, notably on measles, with very interesting scientific results with significant practical consequences. In the present phase of the project the intention is to develop local research capabilities to a point where a local institute will qualify to become partner in an ENRECA-project.

There are many examples of Danida projects drawing on local researchers and research institutions as consultants. A recent and significant example is the role played by the Water Research Institute in Ghana assisting a major Danida project on water supply. Whereas there is no doubt of the usefulness to the project, it may in some cases be relevant to consider the effects on the local research. It is obvious that the high fees paid for consultancies can tempt researchers to strongly reduce the time spent on more genuine research. On the other hand, consultancies give researchers a chance to be exposed to broader problems which may increase the relevance of their research.

Finally, it should be remembered that many aid projects have an innovative character, which means that monitoring and evaluation of progress - or lack of same - provides interesting laboratories for many local researchers, notably within social science.

List of Participants

Hanne Adriansen
Institute of Geography
University of Copenhagen
10, Øster Voldgade
DK - 1350 Copenhagen K

Simon Bolvig
Institute of Geography
University of Copenhagen
10, Øster Voldgade
DK - 1350 Copenhagen K

Mette Bovin
Nordiska Afrikainstitutet
P.O. Box 1703
S-751 47 Uppsala
SWEDEN

Patricia Bussone
I. Krüger Consult A/S
363, Gladsaxevej
DK - 2860 Søborg

Lærke Carlsen
5, Jakob Dannefærdsvej
DK - 1973 Frederiksberg C

Hanne Carus
I.Krüger Consult A/S
363, Gladsaxevej
DK - 2860 Søborg

Jens Christensen
NEAZDP
Danagro Adviser A/S
8, Granskoven
DK - 2600 Glostrup

Per Christian Christensen
40, Sortemosevej
DK - 5892 Gudbjerg

Sofus Christiansen
Institute of Geography
University of Copenhagen
10, Øster Voldgade
DK - 1350 Copenhagen K

Dominique Des Places
c/o Mellemfolkeligt Samvirke
14, Borgergade
DK - 1300 Copenhagen K

Bertil Egerö
PROP (Programmet for
befolkningsfrågor)
Sociologiska Institutionen
University of Lund
Box 114
22100 Lund
SWEDEN

Sidsel Ekås
Udenrigsdepartementet
Postboks 8114 Dep.
0032 Oslo 1
NORWAY

Lars Engberg-Pedersen
Centre of Development Research
5, Gl. Kongevej
DK - 1610 København V

Peter Furu
Danish Bilharziasis Laboratory
1 D, Jægersborg Allé
DK - 2920 Charlottenlund

Tina Svan Hansen
Institute of Geography
University of Copenhagen
10, Øster Voldgade
DK - 1350 Copenhagen K

Ole Jensen
Institute of Geography
University of Copenhagen
10, Øster Voldgade
DK - 1350 Copenhagen K

Søren Kjær
Husdyrbrug og Husdyrsundhed
RUA, 23, Rolighedsvej
DK - 1958 Frb. C.

Holger Koch-Nielsen
17, Højskolevej
Vallekilde
DK - 4534 Hørve

E.O. Kofod
c/o Cowiconsult
15, Parallelvej
DK - 2800 Lyngby

Lars Krogh
Institute of Geography
University of Copenhagen
10, Øster Voldgade
DK - 1350 Copenhagen K

Ole Kubel
Roskilde University Centre
IU, HUS 051
Postbox 260
DK - 4000 Roskilde

Jan Buhl Larsen
Institute of Geography
University of Copenhagen
10, Øster Voldgade
DK - 1350 Copenhagen K

Jonas Erik Lawesson
Växtbiologiska institutionen
14, Villavägen
S-52 36 Uppsala
SWEDEN

Torben Lindkvist
Danida
2, Asiatisk Plads
DK - 1448 Copenhagen K

Michel Lofo
DRPC
B.P. 76
Dori
BURKINA FASO

Christian Lund
Roskilde University Centre
IU, HUS 051
Postbox 260
DK - 4000 Roskilde

H.J. Lundberg
Ministry of Foreign Affairs
2, Asiatisk Plads
DK - 1448 Copenhagen K

Birgitte Markussen
[Department of Ethnography
and Social Anthropology
University of Aarhus]
3, Carl Bernhards Vej
DK - 1817 Frederiksberg C

Siri Melcior-Tellier
Danish Red Cross
27, Blegdamsvej
DK - 2100 Copenhagen Ø

Marlene Meyer
Institute of Geography
University of Copenhagen
10, Øster Voldgade
DK - 1350 Copenhagen K

Jette Michelsen
Ministry of Foreign Affairs
2, Asiatisk Plads
DK - 1448 Copenhagen K

Ivan Nielsen
Institute of Botany
University of Aarhus
68, Nordlandsvej
DK - 8240 Risskov

Trine Louring Nielsen
Institute of Geography
University of Copenhagen
10, Øster Voldgade
DK - 1350 Copenhagen K

Dolf Noppen
Nordic Consulting Group
8, Kirkevej
DK - 2630 Tåstrup

John Olsen
Dept. of Social Anthropology
Moesgård
DK - 8270 Højbjerg

Lise Malling Olsen
11, 4., Dr. Margrethesvej
DK - 8200 Århus N

Reidar Person
SIDA
S 105 25 Stockholm
SWEDEN

Peter Plesner
Ministry of Foreign Affairs, S.8
2, Asiatisk Plads
DK - 1448 Copenhagen K

Gunnar Poulsen
1, Egevænget
Gadevang
DK - 3400 Hillerød

Bjarke Paarup-Laursen
Dept. of Social Anthropology
University of Aarhus
Moesgård
DK - 8270 Højbjerg

Christa Nedergaard Rasmussen
Danish Red Cross
27, Blegdamsvej
P.O. Box 2600
DK - 2100 Copenhagen Ø

Kjeld Rasmussen
Institute of Geography
University of Copenhagen
10, Øster Voldgade
DK - 1350 Copenhagen K

Anette Reenberg
Institute of Geography
University of Copenhagen
10, Øster Voldgade
DK - 1350 Copenhagen K

Inge Schou
Water & Power Planners A/S
48, Rosenvængets Alle
DK - 2100 Copenhagen Ø

Serge Snrech
Club du Sahel
OECD
39-41 Bd. Suchet
75016 Paris
FRANCE

Mike Speirs
Danagro Adviser A/S
8, Granskoven
DK - 2600 Glostrup

Ingeborg Svennevıg
10, 2., Klostergade
8000 Århus C
DENMARK

Leo Stroosnijder
Wageningen Agricultural
University
11, Nieuwe Kanaal
6709 PA Wageningen
THE NETHERLANDS

Aminata D. Traoré
Missira, Rue 20 x 35
Bamako
MALI

Bo Eric Weber
Ministry of Foreign Affairs
Dep. of E & W Africa, S. 8.
2, Asiatisk Plads
DK - 1448 Copenhagen K

Klaus Winkel
Danida
2, Asiatisk Plads
DK - 1448 Copenhagen K

Thorkild Ørum
I. Krüger Consult A/S
363, Gladsaxevej
DK - 2860 Søborg

Participant Profiles

The participants of the workshop were asked to make a short description of the Sahel activities and experience. These summaries are printed below with minor editorial corrections only.

ADRIANSEN, Hanne (Student, Institute of Geography, University of Copenhagen)
Key words: Pastoral nomadism, remote satellite sensing, rangelands management, agricultural systems.
Abstract: Participation in a field course in the Oudalan Province, Burkina Faso in October 1993. A village based study of the agricultural system. As a part of this the main emphasis of my group was to outline the importance of animals in the agricultural strategy. Conflicts between sedentary people and pastoral nomads should be revealed. (Planned studies in Australia in 1994 concerning dryland management and the application of remote sensing in assessments of (semi-) arid rangelands quality.)

BOLWIG, Simon (Student, Institute of Geography, University of Copenhagen)
Key words: Rural socio-economy, production system, Burkina Faso.
Abstract: Being a student of human geography (at the Institute of Geography, University of Copenhagen), my approach to the problems of the Sahel is centred around village and household level studies of agropastoral production systems. From my point of view this approach makes it necessary to incorporate a wide range of social/cultural, economic, and natural factors in the explanation of the state and development of specific production systems. Accordingly, I have an interest in various disciplines, apart from geography, ranging from macroeconomics to anthropology. I also believe that the state and development of production systems must be seen in a longer historical perspective, in order fully to understand the processes at work.

My practical experience of rural Africa limits itself to a case study of ghanaian cocoa producing peasants in 1991, and to a study among Fulani (Rimaiibe) agropastora-lists in the province of Seno in Northeastern Burkina Faso. The latter study took place in the summer of 1993 and will serve as the empirical basis for my masters in geography.

Because the rainy season in the area in 1993 was delayed almost two months, the work of the natural and economic forces that affects the lives of the sahelian people were particularly visible. One conclusion is that in bad years as 1993, the forces of the markets for animals and millet, often deprives the peasants of whatever surplus they may have gathered in good years. Partly as a result of this, is seems that ten years of gold fever in area has had very few positive lasting effects on the economy of the peasants. Another conclusion is that the previous twenty years of low rainfall has had profound effects in terms of the peasant's involvement in the money economy and also very negative consequences for their food self-sufficiency. A third important feature is the decreasing opportunities of migrant work in the neighbouring countries, as a consequence of the general economic crisis. So it is evident that several natural and economic processes work simultaneously, making the lives of the sahelian people extremely difficult.

BOWIN, Mette Sofie (Anthropologist, Nordiska Afrikainstitutet, Uppsala)
Abstract: Has worked for 14 years in West Africa altogether, studying various ethnic groups and inter-ethnic relationships between true pastoralists, agro-pastoralists, and sedentary farmers. Especially in Northern Nigeria, Niger, and Lake Chad Basin. Since 1985 specializing in *pastoral societies* in the arid Sahel: social organization, ecology, and cultural norms of livestock breeders.

Research and information on the following subjects: a) Female roles in Muslim societies south of the Sahara: experience from Republic du Niger, Chad, Cameroon, Mali, Burkina Faso, Nigeria, and Ghana. b) Ethnicity: ethnic relations between peasants, pastoralists, fishermen, etc. in West Africa; ethnic relations in Yugoslavia; and ethnic relations in Sameland - Northern Norway. c) Material culture - analyses, theory, and collection. d) Anthropology and documentary filmmaking (theory and method). e) Pastoral cultures in Africa - from Senegal to Central African Republic, Sahel and Savanna regions (Peul, Wodaabe, Fulbe Borno, etc.). f) Droughts in the Sahelian Zone and strategies for survival.

BUSSONE, Patricia (Sociologist, I. Krüger Consult A/S)
Abstract: Sociologist working in West Africa in the fields of rural water supply, urban water supply, and photovoltaic energy. Experiences from: Cape Verde, Mauritania, Senegal, Gambia, Guinea Bissau, Mali, Burkina Faso, Niger, and Cameroon.

CARLSEN, Lærke (Student, Institute of Anthropology, University of Copenhagen)
Abstract: As a student of Anthropology, University of Copenhagen, I am going to study next year at the University of Ouagadougou. Have been travelling in the Sahel area several times and have been engaged in the Sahel and development issues in my former job as information officer in the Operation Dagsværk campaign. For the last couple of years I have been working in the youth group of Danchurchaid and the Friendship Association Denmark-Burkina Faso.

CARUS, Hanne (Agronomist, I. Krüger Consult A/S)
Key words: Natural resources management, small scale irrigation, agricultural extension in the Sahel and South East Asia.

Abstract: (Burkina Faso): identification of Danida assisted project on natural resource management in the province of Boulgou. The area is characterised by increasing degradation of agricultural and pastoral lands especially due to increasing settlements by immigrants from the central plateau.

(India): agricultural and organisational aspects in relation to appraisals and reviews of Danida assisted programme aiming at improving women's access to agricultural extension services in the four states of Madhya Pradesh, Tamil, Nadu, Orissa, and Karnataka. The programme promotes the training and education of women extension agents at various levels and provides extension services to farm women from small and marginal land holdings.

(Vietnam): water management action plan for the Upper Srepok basin. Covered aspects of land use and of water utilisation in relation to irrigation as part of a comprehensive watershed study.

CHRISTENSEN, Per Christian (Forest Manager)
Key words: Management of natural forests, semi-arid land management, natural resources, sustainable land-use.

Abstract: Several projects in Niger and Burkina Faso have in resent years been comparatively successful in dealing with management of natural resources - especially forests and wood products. In order to make different solutions work, foresters, sociologists, economists, extension workers, and others have developed a collaboration crossing their professional boundaries. The successes are nevertheless limited probably both in time and in space. The projects look more or less sustainable within their own frame and structure, but at best they are isolated islands in a sea of un-sustainability. These islands can, however, serve as benchmarks for the holistic approach necessary for real sustainability.

Essential topics, of which some have only been touched upon briefly in the above mentioned type of projects, are for example: attitudes towards family planning, population growth, quantity versus quality, owners versus users rights of land, change of traditions, and trade relations with the rest of the world. These issues are utmost important to the sustainable solutions of the problems in the Sahelian region, and they are being, and should be, integrated in every approach in any project or programme.

DES PLACES, Dominique (Programme Officer, Mellemfolkeligt Samvirke)
Key words: International Cooperation, personal assistance in Africa. Asia. and Latin America.
Abstract: Mellemfolkeligt Samvirke (MS) is a private association sending development workers to 11 countries, among others 8 in East and Southern Africa. More than 300 development workers on the long term, and 200 workcamp volunteers every year are participating in the work in Eastern and Southern Africa. The main objective of MS programme is double: 1) Partly to contribute to the development of local human resources in the partner country, partly to influence an overall Danish development policy based on solidarity and cross-cultural understanding through information in Denmark. 2) Poverty orientation is the core of MS's strategy and this policy has 4 cornerstones: Development by people, gender orientation, environment and development, and sustainable development.

The main reason for participating in the Sahel Workshop is to gain experiences from West Africa in the field of environment, integrated rural development projects, and programme monitoring.

ENGBERG-PEDERSEN, Lars (Ph.D. Student, Centre of Development Research)
Key words: Natural resource management, decentralization, institutional issues, state - society relations.
Abstract: I am doing a Ph.D. thesis on natural resource management by village councils in Burkina Faso. I have just finished nine months of fieldwork, and have begun to analyze the results.

The key issue to understand is how political processes take place in villages and what the main factors are behind them. The study addresses four groups of factors when trying to explain activities by the village councils created to manage natural resources: 1) environmental degradation; 2) existing (traditional) ways of decision-making; 3) conflicts in and around the village councils; 4) external political and economic factors such as public authorities, projects, NGOs, migration matters, and prices of basic goods.

A preliminary conclusion is that environmental degradation does not influence the work of the councils. Instead, the three other groups of factors are of great importance together with a generally felt wish for infrastructural development in the villages.

FURU, Peter (Environmental Health Biologist, Senior Lecturer, Danish Bilharziasis Laboratory)
Key words: Environmental impact assessment, health impact assessment, environmental management for disease vector control, water resources development.
Abstract - Health Implications of Development Projects: Development projects are designed to confer benefits on the community, including improved standards of living

and health. However, sometimes there are unintended and indirect dis-benefits. These may affect the environment, the socio-economic condition or the health status of some communities. Development resulting in environmental degradation and possibly having negative impacts on human health will be unsustainable and ultimately jeopardize economic progress. Environmental impact assessment (EIA) and health impact assessment (HIA) of development projects are important and necessary management tools making it possible to address potential problems at an early stage in the project planning and design, thereby mitigating or eliminating the environmental and health problems.

In its capacity as a WHO Collaborating Centre for Applied medical Malacology and Schistosomiasis Control and Joint WHO/FAO/UNEP/UNCHS Collaborating Centre for Disease Vector Control in Sustainable Development, the Danish Bilharziasis Laboratory (DBL) is involved in research, training and consultancies on environmental health issues in water resources development projects. The main emphasis is on local capacity building, initiating collaborative research projects on water resources projects, and performing health impact assessments or health risk appraisals of planned or ongoing projects, respectively.

HANSEN, Tina Svan (Student, Institute of Geography, University of Copenhagen)
Key words: Agricultural systems, land use, geographical information systems, remote satellite sensing.
Abstract: Lately, I have participated in a field course in the Oudalan Province, Burkina Faso, October 1993, where it was the intention to make a village based study of the agricultural system. Afterwards, I made a short pre-research in the Krobo area, Ghana in November 1993, which will be followed up by a fieldwork on population pressure versus land use in the area from February to June 1994.

JENSEN, Ole (Student, Institute of Geography, University of Copenhagen)
Key words: Population growth, environmental degradation, migration.
Abstract: Parts of the Sahelian region have long ago been doomed "territoire non viable", territories not able to sustain themselves. A consequence of this has been labour migration from the Sahelian countries to areas with better opportunities for employment. What is the extent and the importance of labour migration? This question was asked by a group of students from the University of Copenhagen during a stay in a village in the Oudalan Province of Burkina Faso in October 1993. We examined the relation between declining productivity in the agricultural sector and the need of looking for employment in other areas in order to support their households, and we tried to estimate the importance of these activities in economic as well as sociological contexts.

KOCH-NIELSEN, Holger (Architect, Civil Engineer, Development Advisory Group)
Key words: Planning, supervision, training, appraisal, and evaluation. Rural and semiurban areas in the Sahel. Natural resources and technology management, technology transfer, simple construction technology, infrastructure, institutional development, training, education approach, community mobilization, operation and maintenance.

KOFOD, E. O. (Forestry Consultant, COWIconsult)
Key words: Project management, agriculture and rural development. Forestry, environment, community mobilization.
Abstract: Recent Sahel based activities: Project manager, Afforestation and Reforestation in the Northern State, Sudan (Danida/UNSO 1988-1993). In this major Sahel project the objective was to contribute to the development of a viable model for introducing desertification combatting measures among the villagers in the Northern State of Sudan. Focus was on awareness raising and mobilization of entire communities. Once the communities were interested in embarking on sand encroachment combatting measures, the project initiated comprehensive training and programmes. Direct expenses were reimbursed by the project but villagers never received compensation of any kind from the project. Food for work project etc. was rejected and in cases in which villagers did not produce their own plants in village nurseries they had to purchase them from project-managed nurseries. Forest trees were thus introduced as marketable commodities thereby encouraging communities or individuals to take up nursery work as income generating activity.

KROGH, Lars (Ph.D. Student, Institute of Geography, University of Copenhagen)
Key words: Soil science, land degradation.
Abstract: Recent projects include: Ph.D study on nutrient circulation and land degradation in Northern Burkina Faso - sustainability of millet production systems, and post doc education at CSIRO Divisions of Soils, Australia.

KUBEL, Ole (Student, Roskilde University Centre)
Key words: Environment and development.
Abstract: My interest lies in the integration of international development and environment. An aspect which I see as very important in the Sahelian area because of the growing population pressure in a harsh natural setting. The efforts of as well national governments and international organizations therefore has to involve the environmental dimension in their development plans.

LARSEN, Jan Buhl (Student, Institute of Geography, University of Copenhagen)
Key words: Migration
Abstract: The traditional Sahelian way of life has become very unstable and unsecured, due to drought, environmental degradation, and population growth. A local response to the changing conditions has been migration in various forms: short/long distance and short/long period.

During a two-week stay in Kolel, a village in Oudalan, Northern Burkina Faso, we (a group of students) experienced that a majority of the families had one or several members outside the village in search of revenues. In this particular village it seemed that the long term migration towards Abidjan (Cote d'Ivoir) had been a very important source of income during several years. There is a kind of rotation system among a group of men, so every man worked a few years in Abidjan, and sent back money for his family in the village. Afterwards he was replaced by another man from the village.

This example shows clearly that it is very important to be aware of how Sahelian peasants are responding to the unstability of agriculture by diversifying their activities. Even though, everybody agreed on the primacy of agriculture, there is no doubt that the village economy is depending on these external revenues.

LAWESSON, Jonas Erik (Assoc. research post doc., Växtbiologiska institutionen, Uppsala)
Key words: Senegal, species-environment relationships, gradients, woody species.
Abstract: Little synthetic work has been done on causal relationships between species, vegetation, and the assumed underlying gradients in West Africa. Most efforts, so far, has been used to describe the flora and vegetation, and rightfully so. With substantial vegetation data starting to accumulate, however, at least in Senegal, the time is due to assess the importance of the different explanatory factors. This is of importance for the formulation of a theory of vegetation dynamics in West Africa, as well as for a proper nature management and sustainable use.

In my study, species-environment relationships in Senegal are revised, using data from 929 plots, 237 woody species, and about 30 explanatory variables, covering location, soils, climate, land-use, nature protection, herbaceous and woody cover, and 49 vegetation types.

The species have been classified into rare, common and very common taxa, as based on their abundance in the data, and analyzed independently. Species and environment correlation's are investigated by means of various multivariate techniques. The explanatory variables are investigated independently of the vegetation types, and forwardly selected through a Monte Carlo perturbation test.

158 species were classified as rare, and some of the variables explaining most of the variance in the data were in the following order: AET/PET ratio, latitude, mean temperature of warmest month, woody cover, longitude.

63 species were classified as common, again with the humidity index (AET/PET) as the most important explanatory variable (0.73 of the total inertia), then followed the presence of Cambisols, Mean annual temperature (TY) and longitude.

16 species were assigned to the "very common" group, and here, the mean annual precipitation explained most of the variance, followed by TY, mean temperature of coldest month (Tmin), the presence of vertisols, and then the presence of some sort of active management.

Further analysis with environmental data, using temperature, precipitation. and hours of sunshine for all 12 months and all 80 sites are in progress.

MARKUSSEN, Birgitte
(Video consultant; Mag.Art. student, Department of Ethnography and Social Anthropology, Aarhus University)
Keywords: Development support communication, educational video production.
Abstract: Experience from consultancies in Northern Nigeria working as educational video consultant covering tree planting campaigns, community participation, and large scale dissemination of video programmes for non-formal education.

I have furthermore carried out several fieldworks in Latin America since 1988 on FAO development support communication projects. My research has focused on the social and cultural innovations evolving around communication projects and communication activities.

MELCHIOR-TELLIER, Siri (Head of Int. Dept./Danish Red Cross)
Key words: population, environment, water supply and sanitation, participation, rural areas, women, training, manuals, monitoring, policies, strategies, and management.
Abstract: Examples of experiences within demographic questions: a) During studies in USA, a study of population policies in francophone Africa. b) Demographic analyses and projections for use by interdisciplinary teams of regional planners. c) Evaluation of census programme of Haiti. d) Evaluation of manpower planning unit of Ghana. e) As UNFPA coordinator in Kabul, developed and monitored programme of UNFPA assistance in Afghanistan, e.g. first population census of country. f) As UNFPA coordinator in Beijing, developed and monitored programme assistance for e.g. census and demographic studies at eleven universities. g) At UNFPA Deputy Director of Africa Branch, statistical/demographic studies in several African countries. h) Today member of background group created by Danida to prepare for the Population Conference 1994.

NIELSEN, Trine Louring (Student, Institute of Geography. University of Copenhagen)
Key words: Pastoral nomadism, remote satellite sensing, rangelands management, agricultural systems.
Abstract: Participation in a field course in the Oudalan Province, Burkina Faso in October 1993. A village based study of the agricultural system. As a part of this the main emphasis of my group was to outline the importance of animals in the agricultural strategy. Conflicts between sedentary people and pastoral nomads should be revealed. (Planned studies in Australia in 1994 concerning dryland management and the application of remote sensing in assessments of (semi-) arid rangelands quality.)

NOPPEN, Dolf (Sociologist, Nordic Consulting Group)
Key words: Project planning, administration, and management.
Abstract: Has since 1987 been working as a consultant, in the fields of project planning, appraisal, management, and evaluation. Sectors covered: water supply and sanitation, rural electrification, community management and participation, co-operative development, natural resources, fisheries, dairy and livestock development.

Of particular relevance for the present workshop, experiences from: Central African Republic, Guinea (Conakry), and Niger with Water Supply and Natural Resources Management. In addition, institutional development experience especially as related to community management and ownership in Uganda, Zambia, and Malawi.

OLSEN, John (Agroforester, Department of Ethnography and Social Anthropology, Aarhus University)
Abstract: At present I am employed at the above mentioned institute doing research on: Peoples strategies in utilising cultivated and uncultivated resources in Africa with special attention to woody vegetation and a focus on the *Faidherbia albida* agri-silvopastoral system in Africa.

The *Faidherbia albida* farming system is prevalent in the Sahelian zone, both heavily degraded systems and well traditionally managed systems and it is considered an asset by farming communities. Scientists see it as an alternative to shifting cultivation. Where *Faidherbia albida* is incorporated into the farming systems approximately twice the human population can be supported. It is important to find the socio-economic rationale for why apparently successful management practices are abandoned and why they in other cases persist.

PAARUP-LAURSEN, Bjarke (Lecturer, Department of Ethnography and Social Anthropology, Aarhus University)
Key words: Social and cultural studies in West Africa, rural development in West Africa, primary health care in the Third World, the relationship between traditional and modern health care systems, cultural analysis, ritual and symbolic studies.

Abstract: During the last ten years I have carried out a number of long term fieldworks in Africa, mainly Northern Nigeria. The first projects focused on general social and cultural analysis, especial ritual and symbolism, while the last project was an interdisciplinary medical-anthropological study of the use of health care systems in a number of societies in Northern Nigeria.

I am presently employed at the Department of Ethnography and Social Anthropology, Aarhus University. The department is presently focusing its research on the relationship between society and nature, and I am presently developing research within this field.

RASMUSSEN, Christa Nedergaard (Project consultant/Danish Red Cross)
Key words: environment, education, rural areas, children, youth.
Abstract: Environmental Education projects for children in the primary schools and for young people outside the school system. Coordinate projects in: Senegal, Burkina Faso, and Sudan.

RASMUSSEN, Kjeld (Associate professor, Institute of Geography, University of Copenhagen)
Key words: Agricultural systems, natural resources, remote sensing.
Abstract: Understanding of long-term changes in agricultural systems and land use and in natural resources over large areas requires information which can only be obtained using remote sensing methods. Information extracted from remote sensing data provides inputs to scientific studies as well as to natural resource management on village- to continental scale. Financed by Danida, the Institute of Geography, University of Copenhagen (IGUC), collaborates with 'Centre de Suivi Ecologique', Dakar, Senegal, on the development of methodologies and software, allowing simple low-cost nation-wide monitoring of a number of agricultural and environmental themes and parameters, such as: crop yield, natural grassland productivity, rainfall and evapotranspiration, bush fires, forest cover, and agricultural land use.

The methodologies developed are implemented in software, widely used in Africa and elsewhere, for environmental applications of remote sensing data.

In a broader perspective, research activities at IGUC aim at integrating remote sensing and other spatial information on natural resources in the study of agricultural systems and natural resource management in the Sahel, and at understanding the interrelationships between land use/cover and climate on a regional scale.

REENBERG, Anette (Associate professor, Institute of Geography, University of Copenhagen)
Key words: Agricultural systems, land-use dynamics, monitoring by remote sensing.
Abstract: Recent project activities include: a) investigations of millet production systems in Northern Burkina Faso with focus on changes in land use, cultivation intensity, nutrient circulation (in cooperation with Lars Krogh); b) natural resource mapping and landscape dynamics in Niger/Diffa (1992); c) responsible for field course in Northern Burkina Faso focusing on village based studies on agricultural systems.

SVENNEVIG, Ingeborg (M.A. student, Department of Ethnography and Social Anthropology, Aarhus University)
Key words: Resource utilisation, land registration, land tenure, representatives in local communities.
Abstract: That land management and 'overpopulation' are coherent factors became clear to me during a fieldstudy in Luangwa Valley, Zambia and a two-week consultancy in Ngamiland and Chobe District, Botswana. Both areas are rich on wildlife and wilderness.

Despite the low population densities in these semi-arid areas, the contradictions between human settlement and utilisation of natural resources and the preservation of wildlife have led managers to define the problem as an 'overpopulation' of a potentially wealthy wilderness. Hence, land management has meant to restrict the settlement areas and control the land utilisation of the human inhabitants.

In this context two significant issues appears considering the specific relationships between human populations and potential environmental degradation. 1) 'Overpopulation' will often exist as an imbalance between population size and available land and the capacity of it. Hence, I focus on the available *area* for the human population concerned, and the adaption of specific utilisation methods to specific types of territories.

2) 'Overpopulation' can be reformulated as 'un-suitability of human activities', both as too many children, and as futile use of natural resources. I have studied the question of *representatives* in the human community, both as a means to understanding the existing management of resources, and, ideally, as a dynamic factor in the relationship between the 'overpopulation' and external prescriptions, such as family planning and restrictions on hunting rights.

Presently, I am identifying the anthropological tools, both as existing analysis and as general theories, appropriate for conduction research on these issues.

REPORTS FROM THE BOTANICAL INSTITUTE,
UNIVERSITY OF AARHUS
1. **B. Riemann:** Studies on the Biomass of the Phytoplankton. 1976. Out of print.
2. **B. Løjtnant & E. Worsøe:** Foreløbig status over den danske flora. 1977. Out of print.
3. **A. Jensen & C. Helweg Ovesen (Eds.):** Drift og pleje af våde områder i de nordiske lande. 1977. 190 p. Out of print.
4. **B. Øllgaard & H. Balslev:** Report on the 3rd Danish Botanical Expedition to Ecuador. 1979. 141 p. Out of print.
5. **J. Brandbyge & E. Azanza:** Report on the 5th and 7th Danish-Ecuadorean Botanical Expeditions. 1982. 138 p.
6. **J. Jaramillo-A. & F. Coello-H.:** Reporte del Trabajo de Campo, Ecuador 1977—1981. 1982. 94 p.
7. **K. Andreasen, M. Søndergaard & H.-H. Schierup:** En karakteristik af forureningstilstanden i Søbygård Sø — samt en undersøgelse af forskellige restaureringsmetoders anvendelighed til en begrænsning af den interne belastning. 1984. 164 p.
8. **K. Henriksen (Ed.):** 12th Nordic Symposium on Sediments. 1984. 124 p.
9. **L. B. Holm-Nielsen, B. Øllgaard & U. Molau (Eds.):** Scandinavian Botanical Research in Ecuador. 1984. 83 p.
10. **K. Larsen & P. J. Maudsley (Eds.):** Proceedings. First International Conference. European-Mediterranean Division of the international Association of Botanic Gardens. Nancy 1984. 1985. 90 p.
11. **E. Bravo-Velasquez & H. Balslev:** Dinámica y adaptaciones de las plantas vasculares de dos ciénagas tropicales en Ecuador. 1985. 50 p.
12. **P. Mena & H. Balslev:** Comparación entre la Vegetación de los Páramos y el Cinturón Afroalpino. 1986. 54 p.
13. **J. Brandbyge & L. B. Holm-Nielsen:** Reforestation of the High Andes with Local Species. 1986. 106 p.
14. **P. Frost-Olsen & L. B. Holm-Nielsen:** A Brief Introduction to the AAU - Flora of Ecuador Information System. 1986. 39 p.
15. **B. Øllgaard & U. Molau (Eds.):** Current Scandinavian Botanical Research in Ecuador. 1986. 86 p.
16. **J. E. Lawesson, H. Adsersen & P. Bentley:** An Updated and Annotated Check List of the Vascular Plants of the Galapagos Islands. 1987. 74 p.
17. **K. Larsen:** Botany in Aarhus 1963 - 1988. 1988. 92 p.

AAU REPORTS:
18. Tropical Forests: Botanical Dynamics, Speciation, and Diversity. Abstracts of the AAU 25th Anniversary Symposium. Edited by **F. Skov & A. Barfod.** 1988. 46 pp.
19. Sahel Workshop 1989. University of Aarhus. Edited by **K. Tybirk, J. E. Lawesson & I. Nielsen.** 1989.
20. Sinopsis de las Palmeras de Bolivia. By **H. Balslev & M. Moraes.** 1989. 107 pp.
21. Nordiske Brombær (Rubus sect. Rubus, sect. Corylifolii og sect. sect. Caesii). By **A. Pedersen & J. C. Schou.** 1989. 216 pp.
22. Estudios Botánicos en la "Reserva ENDESA" Pichincha - Ecuador. Editado por **P. M. Jørgensen & C. Ulloa U.** 1989. 138 pp.
23. Ecuadorean Palms for Agroforestry. By **H. Borgtoft Pedersen & H. Balslev.** 1990. 120 pp
24. Flowering Plants of Amazonian Ecuador - a checklist. By **S. S. Renner, H. Balslev & L. B. Holm-Nielsen,** 1990. 220 pp.
25. Nordic Botanical Research in Andes and Western Amazonia. Edited by **S. Lægaard & F. Borchsenius,** 1990. 88 pp.
26. HyperTaxonomy - a computer tool for revisional work. By **F. Skov,** 1990. 75 pp.

27. Regeneration of Woody Legumes in Sahel. By **K. Tybirk**, 1991. 81 pp.
28. Régénération des Légumineuses ligneuses du Sahel. By **K. Tybirk**, 1991. 86 pp.
29. Sustainable Development in Sahel. Edited by **A. M. Lykke, K. Tybirk & A. Jørgensen**, 1992. 132 pp.
30. Arboles y Arbustos de los Andes del Ecuador. By **C. Ulloa Ulloa & P. M. Jørgensen**, 1992. 264 pp.
31. Neotropical Montane Forests. Biodiversity and Conservation. Abstracts from a Symposium held at The New York Botanical Garden, June 21–26, 1993. Edited by **Henrik Balslev**, 1993, 110 pp.
32. THE SAHEL: Population. Integrated Rural Development Projects. Research Components in Development Projects. Proceedings of the 6th Danish Sahel Workshop, 6—8 January 1994. Edited by **Annette Reenberg & Birgitte Markussen**. 1994. 171 pp.
33. The Vegetation of *Delta Du Saloum* National Park, Senegal. By **A. M. Lykke**, 1994 (in press). Ca. 90 pp.

ORDER FORM

To Aarhus University Press
Aarhus University
DK-8000 Aarhus, DENMARK
Fax (+45) 8619 8433

I would like to order the following issues of AAU REPORTS:

_____

I enclose the equivalent of 80 DKr per issue (13 USD).
Residents of the EU should add 25% Danish VAT.

Please send the books to:

Name:
Street:
Town:
Country: